Sustainable Automotive Technologies 2010

Jörg Wellnitz · Aleksandar Subic · Martin Leary
Editors

Sustainable Automotive Technologies 2010

Proceedings of the 2nd International Conference

Springer

Editors
Professor Dr.-Ing. Jörg Wellnitz
Private Institute of Technology
and Artistic Design (ITD e.V.)
Marie-Curie-Straße 6
85055 Ingolstadt, Germany
joerg.wellnitz@fh-ingolstadt.de

Professor Dr. Aleksandar Subic
RMIT University, School of Aerospace
Mechanical & Manufacturing Engineering
P.O.Box 71
Bundoora, Victoria 3083, Australia
Aleksandar.Subic@rmit.edu.au

Dr. Martin Leary
RMIT University, School of Aerospace
Mechanical & Manufacturing Engineering
P.O.Box 71
Bundoora, Victoria 3083, Australia
martin.leary@rmit.edu.au

ISBN 978-3-642-10796-2 e-ISBN 978-3-642-10798-6
DOI 10.1007/978-3-642-10798-6
Springer Heidelberg Dordrecht London New York

Library of Congress Control Number: 2010920099

Cover design: WMXDesign GmbH, Heidelberg

Printed on acid-free paper

Springer is part of Springer Science+Business Media (www.springer.com)

PREFACE

Mobility is an essential part of our lives. The ability to move freely is central to meeting our social and economic needs. For this reason we have embraced the car over the past century perhaps more than any other technology or consumer product. Today there are around 900 million vehicles on the world`s roads with another 60,000,000 new vehicles produced each year worldwide. The scale of the automotive industry is significant and far reaching. For example, it is estimated that around two thirds of world`s oil output goes to transportation whereby road vehicles alone consume around 40% of the world`s rubber and 25% of the world`s glass, with the consumption of raw materials and other resources further growing due to the rapid development of the automotive sector in China, India, Thailand and Mexico. Transportation accounts for around 25% of greenhouse emissions worldwide, whereby 90% of transport related emissions come from road vehicles, predominantly cars. Clearly, current levels of consumption and emissions are unsustainable. This in turn suggests that mobility as we know it, based on the traditional vehicle technology and existing production and consumer practices, is unsustainable.

The challenge of developing new sustainable approaches to mobility confronts industries and our societies in general. The concept of sustainable mobility is multidimensional and the challenge of achieving it is quite complex. Based on current knowledge it is becoming painfully clear that there is no "sliver bullet" or single technology available at present to address this challenge. To succeed we will most likely have to pursue a range of different technologies and approaches with short-term and long-term gains. This book aims to draw special attention to the research and practice focused on new technologies and approaches capable of meeting the challenges to sustainable mobility. It comprises of selected international peer reviewed papers presented at the 2. International Conference on Sustainable Automotive Technologies held in Gut Ising from 23-26 February 2010. The research here focuses primarily on radical technical advancements and approaches that are able to meet social, economic and environmental targets in both local and global contexts.

The book has been edited and its contents divided into the following interrelated chapters:

- Technologies and infrastructure for sustainable mobility
- Solution for reduced pollution and congestion
- Sustainable vehicle design and manufacture
- New vehicle concepts
- New structures and materials
- New power train technologies
- End-of-life vehicle management
- Vehicle and public safety

We gratefully acknowledge the authors and the referees who have made this publication possible with their research work and written contributions. We would also like to thank the IFG Ingolstadt, Audi AG Ingolstadt, RMIT University and University of applied Sciences, Ingolstadt for their generous sully, we hope that a book on the multidisciplinary subject of sustainable mobility, as diverse in topics and approaches as this one, will be of interest to automotive technology researchers, policy makers, practitioners and enthusiasts whatever their background or persuasion.

Our special thanks also to the Mr. Sebastian Lorenz, who was in charge of the editorial compilation of the book.

We are also very thankful to Mrs. Birgit Paolini, in the function as the secretary of the conference, both doing a great job for the ICSAT 2010.

Aleksandar Subic, Martin Leary and Jörg Wellnitz
Editors
February 2010

TABLE OF CONTENTS

Part 1
Technologies and infrastructure for sustainable mobility

TOWARDS A SUSTAINABLE VEHICLE DEVELOPMENT

W.-P. Schmidt

Ford Werke GmbH, Henry Ford Str., D-E479/W03, D-50735 Koeln, Germany
E-mail: wschmi18@ford.com

Abstract Sustainable automotive technologies need to be developed following holistic assessment criteria and need to be embedded in an integrated approach. The methods developed over time broadened the aspects covered and considerably changed the focus. Design-for-Disassembly and Design for Recycling focused on the end-of-life vehicle. Later, Design-for-Environment considered all relevant environmental aspects. Finally, Ford of Europe's Product Sustainability Index (PSI) is a sustainability management tool for product development management considering environmental and economic life cycle aspects, as well as safety, mobility capability and key environmental aspects. Based on such a broad assessment basis a multitude of vehicle technologies are promising solutions.

1. The sustainability challenge

Given its daily and dominant appearance in modern world, automotive mobility had been criticized for decades in particular due to environmental issues.

The consumption of oil and resources has raised concerns about limits to further growth (Meadows et. al. 1972), concerns that came back with more or less questioned statements about oil peaking (Cavallo 2004) in recent time accompanied with highest oil prices until recession hit the world economy in 2008. End-of-Life Vehicles and the recycling of non-metals led to several regulations in particular in Europe in the 90es followed now in Asian regions although vehicles have been and are one of the most recycled, complex products.

While new vehicles tailpipe emissions of NOx will be aggressively decreased by a factor of around 20 in Europe (1993–2015) their specific CO2 emissions are still seen as unacceptable high in spite of a reduction of 30% (1995–2015).

This view is supported by the alarming scientific projections suggesting a continued climate change with severe consequence for world economies, humans and their environment (IPPC 2007). The consensus is that the average temperature should not increase by more than 2°C.

Nevertheless, the substitution of old vehicles by new vehicles as forced by recent fleet renewal programs (cash-for-clunker etc.) showed clear environmental

J. Wellnitz et al. (eds.), *Sustainable Automotive Technologies 2010*,
DOI 10.1007/978-3-642-10798-6_1, © Springer-Verlag Berlin Heidelberg 2010

benefits – in Germany e.g. on average CO_2 by 20%, NO_x by 87%, Particles by 99% and reduced drive-by-exterior noise.

However, beyond environmental aspects, societal ones are also part of sustainability. The societal benefits of automotive mobility are the basis of its global success based on offering individual mobility also measurable in terms of number of passengers and size of luggage compartment. Of course there are also societal challenges. The main challenge being safety for passengers and pedestrians.

Last but not least there are economical aspects of sustainability. Mobility that is best rated from an environmental and societal point of view is only reasonable if affordable.

2. Initial thinking: Design-for-Disassembly (DfD) and Design-for-Recycling (DfR)

Automotive engineering respecting sustainability aspects focused in the 70's to the beginning of the 90's on the issue of end-of-life vehicles. Re-use and recycling of products have been seen as a fundamental pre-requisite of a sustainable approach for products. This would apply also to the non-metallic content of vehicles. Without closing the material loops the danger of limiting growth has been seen. Re-use and *recycling* have been seen *as a target in itself.* To ensure good re-use and recycling, it has been seen as a task for product development to design automotive parts in a way that dismantling of targeted parts would be as easy as possible (e.g. part accessibility, type and number of fasteners enable easy dismantling). Material selection has been suggested to consider recycling aspects (e.g. material compatibility, reduced material diversity) and the use of recycled materials should increase demand for recycling products.

In consequence design constraints for a Design-for-Disassembly (DfD) and Design for Recycling (DfR) are very severe and might be in conflict with other vehicle attributes (craftsmanship, weight, packaging, etc.) and customer demands (e.g. soft-touch surfaces).

In addition, doubts about such an approach increased as there is a fundamental difference between vehicle assembly (workers trained to assemble few parts for one or two vehicles) and disassembly (mixture of hundreds of vehicle types, aged and damaged vehicles or screws etc.). Real-world dismantling tests showed that therefore, only a small portion of the dismantling time can be addressed by product design at all. Most theoretical linkages between design aspects as length of screws, accessibility, visibility etc. are overruled in practice by questions as operators experience and work organization (Kazmierczak et al. 2005; Kriegl et al. 2005). This means, design actions (DfD) do often turn out as being not effective. This is in particular relevant when looking at results showing that also from an environmental perspective dismantling is not clearly preferably compared to post-shredder treatment separation (Krinke 2005).

Also the DfR approach has been questioned by a series of studies. From a purely environmental perspective, when only focusing on the end-of-life phase there is no evidence that mechanical recycling is environmentally favorable compared to other ELV treatment options for plastic waste (Jenseits et al. 2003; Nürrenbach 2003; Krinke 2005). From a purely environmental perspective - when taking the whole vehicles life cycle into account - the end-of-life phase of certain kind of non-metals does not play any significant role in terms of potential environmental impacts or recycling credits, i.e. all these efforts result in no remarkable improvement for the environment (Castro et al. 2003; Schmidt et al. 2004; Reuter et al. 2006). Recycling-driven changes to the product design can sometimes jeopardize the overall environmental vehicle performance (e.g. through material selection potentially leading to increased weight like efforts to minimize material complexity (Schmidt et al. 2004). The overall environmental performance can sometimes been jeopardized when sub-optimizing the ELV treatment by changes in the product development (e.g. material selection regarding lightweighting) (Nürrenbach 2003; Schmidt et al. 2004; Reuter et al. 2006).

3. Next step: Design for Environment (DfE)

The knowledge about the holistic aspects covering the whole (vehicle) life cycle led to a change towards integrating environmentally life cycle thinking into product development. This was documented in ISO TR 14062 – a technical report about 'integrating environmental aspects into Product design and development'. It is recommended to transform the general guidance of ISO TR 14062 to the specifics of a company and product (Quella and Schmidt 2003).

In consequence there are numerous DfE tools (e.g. Multi-Criteria-Requirement Matrix, Environment-Failure Mode Effect Analysis (Schmidt 2001), checklists, simplified life cycle tools or comprehensive material selection tools (Rebitzer and Schmidt 2003). For all these tools the holistic approach is important – covering not only one but all important environmental impacts relevant for the product.

However, it turned out that also this perspective was unsatisfactory. Products that are designed for low environmental impact – or even better: positive environmental impacts – are not in any case also sustainable. For example the 'greenest' product will fail if it is not affordable. Step-by-step aspects beyond environmental issues needed to be included in the tools.

4. Today: Design-for-Sustainability (DfS)

The first logical step was the extension of the environmental life cycle tools to cover also economic aspects. There Life Cycle Costing is an ideal tool (Schmidt 2003). However, a more comprehensive approach was needed to reflect the key vehicle attributes contributing to all dimensions of sustainability. In consequence, Ford Motor Company Europe introduced a new product design management tool called Product Sustainability Index (PSI), by which various dimensions of sustainability have been combined into a comprehensive set of metrics for steering vehicle development. PSI reflects the overall key impact of the different vehicle attributes and makes the trade-offs visible (*e.g.* between lifecycle global warming potential and the lifecycle cost of ownership). The resulting PSI indicators are:

1. Lifecycle global warming potential (greenhouse emissions along the lifecycle)
2. Lifecycle air quality potential (summer smog creation potential (POCP) along the lifecycle (VOCs, NO_x))
3. Sustainable materials (use of recycled and natural materials)
4. Restricted substances
5. Drive-by-exterior noise
6. Safety (pedestrian and occupant)
7. Mobility capability (luggage compartment volume plus weighted number of seats related to vehicle size)
8. Lifecycle ownership costs (vehicle price plus 3-year fuel costs, maintenance costs, taxation and insurance minus residual value)

The implementation of the PSI has been done in a process-driven approach. Clarification of the organizational context is of utmost importance in such complex, big corporations to make the individual departments directly responsible for that specific aspect of sustainability that can be impacted by their area of responsibility. The main affected departments include product development and manufacturing but also human resources and external affairs (Schmidt 2008).

5. Sustainable vehicle technologies

Based on these sustainability criteria a broad set of vehicle technologies seems to be promising along the time line. For the next years gasoline and diesel engine technologies are most promising to balance demands for low CO2 along the life cycle and affordability. This can be achieved by down-sizing, super-charging and other fuel efficiency technologies. Already without any electrification the Ford Fiesta Econetic lowered emissions to 98 g CO2/km (2009). Adding start-stop resulted for the bigger Ford Focus Econetic in also only 99 g CO2/km. Electrification may be needed to achieve lower emissions (from micro-hybrid to full electric vehicles). A higher share of electrification leads to lower CO2 emissions but also to lower affordability. This needs to be balanced to keep these solutions really sustainable.

It should be noted that electric vehicles are also along the life cycle truly zero-CO_2 vehicles – at least in Europe: Currently non-electric vehicles are not part of the European Emission Trading Scheme (EU ETS), i.e. they are subject to efficiency regulations (targets for the fleet g CO_2/km) but have no absolute emission cap (total maximum CO_2 emissions). If a vehicle is now fully electric, it leaves the non-EU ETS sector as the electricity used is covered in EU ETS. EU ETS is capping the maximum emissions of all sectors in this system – including electricity production. Any additional electricity demand due to electric vehicles is lowering the gasoline or diesel combustion but is not increasing the CO_2 emissions in the (capped) EU ETS sector. Thus they are zero CO_2.

Alternative fuel vehicles are other interesting vehicle technologies. For electric vehicles as well as alternative fuel vehicles also other stakeholders are needed to ensure a sustainable success of these vehicles: the infrastructure offering fuel as well as electric charging points needs to be established to ensure that these technologies have a sustainable future. In addition, tax incentives need to kick-off the demand.

References

Castro M.B.G., Remmerswaal J.A.M., Reuter M.A.: Life Cycle Impact Assessment of the Average Passenger Vehicle in the Netherlands. In: IJLCA 8 (5) 297 – 304 (2003)

Cavallo A. J.: Hubbert's Petroleum Production Model: An Evaluation and Implications for World Oil Production Forecasts, *Natural Resources Research*, Vol. 13, No. 4, 211-221, December 2004

Intergovernmental Panel on Climate Change (IPCC), Pachauri, R.K., Reisinger, A. (Editors) (2007): Climate Change 2007 – Synthesis Report. Geneva

Jenseit W, Stahl H, Wollny V, Wittlinger R (2003): Recovery Options for Plastic Parts from End- of- Life Vehicles: an Eco- Efficiency Assessment; Öko- Institut: Darmstadt.

Kazmierczak K., Mathiassen S.E., Forsman M., Winkel J: An integrated analysis of ergonomics and time consumption in Swedish 'craft-type 'car disassembly. In: Applied Ergonomics 36 (2005) pp 263-273

Kriegl M., Poxhofer R., Schmidt W.-P., Alonso JC: Dismantling Manuals for EES. (2005), Dismantling Manuals for EES www.sees-project.net/contents/File/D3_Report.pdf

Krinke S: Life Cycle Assessment of the Volkswagen-Sicon Process. SETAC Europe Conference Lille 23rd May 2005

Meadows D. H, Meadows D. L., Randers J, Behrens W. W. (1972): *The Limits to Growth*. Universe Books, New York

Nürrenbach T (2003): Recovery of plastic parts from ELV' s – Evaluation of environmental impacts according to the LCA method and economic analysis; Fraunhofer Institut für Verfahrenstechnik, Freising.

Quella, F.; Schmidt, W.-P.: Integrating Environmental Aspects into Product Design and Development. The new ISO TR 14062. In: International Journal of LCA 8 (2), Landsberg: ecomed publishers (2003)

Rebitzer G., Schmidt W.-P.: Design for Environment in the Automotive Sector with the Materials Selection Tool euroMat In: Gate to EHS: Life Cycle Management – Design for Environment, March 17th , 2003 pp 1-4. Landsberg: ecomed publishers (2003)

Reuter, M.A., van Schaik A., Ignatenko A., de Haan G.J: Fundamental limits for the recycling of end-of-life vehicles. In: Minerals Engineering (Elsevier Science), pp 433-449 (2006)

Schmidt W.-P., Dahlqvist E., Finkbeiner M., Krinke S., Lazzari S., Oschmann D., Pichon S., Thiel Ch.: Life Cycle Assessment of Lightweight and End-of-Life Scenarios for Generic Compact Class Passenger Vehicles. In: International Journal of LCA 9 (6), pp 405 – 416. Landsberg: ecomed publishers (2004)

Schmidt W.-P.: Life Cycle Costing as Part of Design for Environment – Environmental Business Cases. In: International Journal of LCA 8 (3), pp 167-174. Landsberg: ecomed publishers (2003)

Schmidt, W.-P.: Developing a Product Sustainability Index. In: Measuring Sustainable Production. OECD Sustainable Development Studies, pp 115 – 126. OECD: Paris, 2008

Schmidt, W.-P.: Umwelt-Fehlermöglichkeiten- und Einfluss-Analyse. In: DIN Deutsches Institut für Normung e.V. (Hrsg.) Umweltgerechte Produktentwicklung – Ein Leitfaden für Entwicklung und Konstruktion; S. 1-12. Beuth Verlag: Berlin, Wien Zürich, März 2001

EADS COMMITMENT TOWARDS SUSTAINABLE MOBILITY

A. Godula-Jopek

EADS Innovation Works, TCC 6 Energy & Propulsion,
Willy Messerschmitt Str., 81663 Munich, Germany
E-mail: agata.godula-jopek@eads.net

Abstract Innovation is a permanent driver for EADS and as such the company objective is to meet the Vision for 2020 of the Advisory Council for Aeronautic Research in Europe (ACARE) goals for a more environmentally friendly air transport. For EADS, anticipating and addressing the environmental challenges is already a major driver and a clear technological, industrial and managerial choice. Developing a portfolio of technologies to satisfy future customer/product requirements, with close interlinking of mobility and sustainability is the key element in the EADS strategy.

1. Introduction

EADS is a global leader in aerospace, defence and related services. The Group includes Airbus as the leading manufacturer of commercial aircraft, with Airbus Military covering tanker, transport and mission aircraft, Eurocopter as the world's largest helicopter supplier and EADS Astrium, the European leader in space programmes from Ariane to Galileo. Its Defence & Security Division is a provider of comprehensive systems solutions and makes EADS the major partner in the Eurofighter consortium as well as a stakeholder in the missile systems provider MBDA.

2. Global goals for environment

2.1 ACARE goal 2020

ACARE (Advisory Council for Aeronautics Research in Europe) established new standards in commercial aviation. Commitments of the European Aerospace Industry with Visionary Goals for 2020 in terms of environment, relative to 2000, are as follows:

J. Wellnitz et al. (eds.), *Sustainable Automotive Technologies 2010*,
DOI 10.1007/978-3-642-10798-6_2, © Springer-Verlag Berlin Heidelberg 2010

- Noise Reduction by 50% (on a cumulative basis)
- CO_2/Fuel Consumption Reduction by One Half, thereof:
  - 15–20 % Power plant
  - 20–25 % Airframe
  - 5–10 % Air Traffic Management
- NO_x Reduction by 80 %

The goals are planned to be reached through new engine concepts, cycle optimization and technology implementation. In addition, focused on process improvements, a substantial reduction of operating costs, five-fold reduction of accident rates, cutting in half the time to entry into service and improved synergies between civil and military domains is mandatory. "Clean Sky" Joint Technology Initiative for Aeronautics & Air Transport will focus on developing technologies that demonstrate and validate the breakthroughs necessary to make major steps towards the goals sets by ACARE and REACH (Directive deals with the Registration, Evaluation, Authorisation and Restriction of Chemical substances).

2.2 IATA goal 2017

Detailed investigations are required to assess the potential of revolutionary technologies expected after 2020 for approaching a zero-emissions future for aviation. IATA (International Air Transport Organisation) has set a goal of 10 % alternative fuels of non-crude oil sources by 2017. This will be depending on developments in the aviation and fuels industry. Sustainable biofuels have the potential to reducing GHG (greenhouse gas) emissions across the whole lifecycle. To assure that biofuels have a lower GHG footprint, production must meet key sustainability criteria. The certification of a blend of hydro processed renewable jet fuel is currently targeted for end of 2010 and a 100 % renewable jet fuel certification by 2013 (IATA 2008 Report on Alternative Fuels).

IATA identified a set of specific actions that should be undertaken by the aviation industry to reduce fuel burn and thus greenhouse gas emissions, keep the airlines economically viable, and ultimately define an industry that has a net zero carbon contribution to the global atmosphere. Fig.1. presents a projection of aviation fuel burn and CO_2 emissions per revenue ton kilometre (RTK), relative to 2005 value.

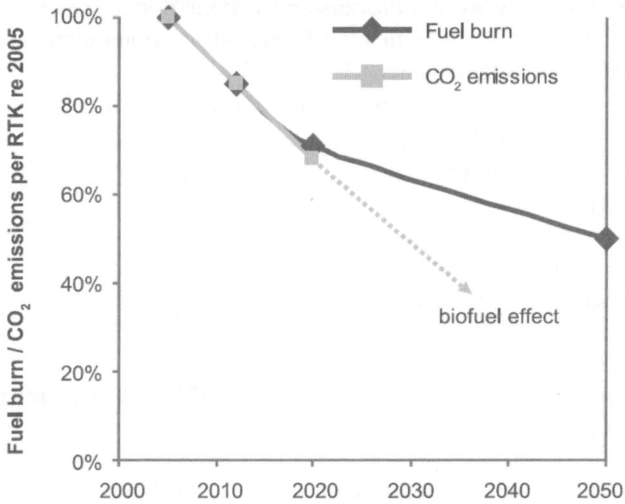

Fig. 1 Projection of aviation fuel burn and CO_2 emissions per revenue ton kilometre (RTK). Source: IATA Technology Roadmap Report 3rd edition June 2009.

3. Importance of the innovation in EADS

EADS must keep the innovation pipeline constantly full in order to replace ageing technologies and/or processes. R&T moving towards being an eco-efficient enterprise are focusing on green energy and propulsion (see also section 3.2.1).

3.1 Vision 2020

EADS objective is to meet the Vision for 2020 of the ACARE goals for more environmentally friendly air transport. This is being realised by moving towards an eco-efficient company and acting consistently to sustain future development. As stated before, for EADS, anticipating and addressing the environmental challenges is already a major driver and a clear technological, industrial and managerial choice.

3.1.1 EADS environmental research initiatives

EADS has set up several research initiatives in order to address the ACARE goals for more environmentally friendly air transport:

- Ramp-up of the Energy and Propulsion technical domain
- Greener helicopter technologies (Le Bourget Bluecopter demonstrator)
- Advanced Propulsion Systems
- Fuel Cells, Fuel Processing, Hydrogen and Hydrogen Storage
- Roadmaps for alternative fuels
- SESAR – Single European Sky Air Traffic Management Research & trajectory optimized operation
- Noise and vibration reduction
- Increased aerodynamic efficiency
- More effective tools for the optimization of power management

3.1.2 Bluecopter concept as demonstration of EADS' commitment to sustainable aviation – towards cleaner sky

The Bluecopter technology demonstrator (Fig. 2) has been presented by EADS Innovation Works at Paris Airshow, at Le Bourget in 2009. The Bluecopter technology demonstrator represents part of EADS' participation in the European CleanSky Joint Technology Initiative research towards achieving the ACARE goals, where Eurocopter and EADS Innovation Works cooperate with other partners to reduce the specific fuel consumption by over 30 %, decrease CO_2 emissions by up to 40 % and reduce NO_x emissions by up to 53 %. It was highlighted by EADS CTO, Dr. Jean Botti that "Bluecopter features technologies which may pave the way for future development and trigger innovation towards the targeted emission reductions".

Fig. 2 Bluecopter is the continuous improvement endeavor to reduce the environmental footprint of rotorcraft through their entire lifecycle, from initial design to effective end-of-life recycling and noise reduction.

Demonstrator Bluecopter (concept):

- Helicopter Mock Up with integrated high pressure combustion piston engine
- Engine equipped with state-of-the-art exhaust gas purification
- Engine capable of running to demonstrate function of exhaust gas purification and low fuel consumption

3.2 EADS innovation works

EADS Innovation Works (EADS IW), as a transnational network of research centres working for EADS Business Units (BUs) is targeting on identifying new technologies that create value and develop them up to TRL 3 and guarantying the technical innovation potential for EADS up to TRL 6. The mission of Innovation Works is to optimize the emergence of new technologies, promote the development of value creating products and services, and support the development of product demonstrators. Main customers of EADS IW are EADS divisions: Airbus including Airbus Military, Eurocopter, EADS Astrium, Defence & Security. EADS IW is based on seven Technical Capability Centres (TCCs), with wide spectrum of competencies in Composites Technologies, Metallic Technologies & Surface Engineering, Structures Engineering, Production & Aeromechanics, Sensors, Electronics & Systems Integration, Simulation, IT & Systems Engineering, Energy & Propulsion and Innovative Concepts.

3.2.1 TCC 6 Energy and Propulsion

Technical Capability Centre 6 (TCC 6), Energy & Propulsion is focused on meeting the future energy and propulsion needs of the EADS group. Moving towards realization of set ACARE goals, TCC 6 is investigating solutions for **green energy and propulsion**:

- Key focus on integration of alternative power generation technologies, including Fuel Cells, and novel Energy Storage
- Driving Greener product specifications in energy efficiency, emissions, and noise
- Strong involvement with the development of alternative fuels, including biofuels
- Integrates mechanical, electronic and systems engineering to provide optimised solutions for energy efficiency and power management

4. Axes of development

4.1 Architecture level (more electric aircraft)

There is a clear trend visible towards more electrical power in general industry and aerospace – More Electric Aircraft, eliminating hydraulics and pneumatics and increasing of electric functions. The expected benefits are weight saving on aircraft level, improved efficiency and maintenance (e.g. elimination of potential hydraulics leakage sources), dissimilarity of power sources, more flexible reconfiguration in case of power generation failure, enable Green Aircraft.

4.2 Component level

On the component level, the potential integration of fuel cells, batteries and supercapacitors into EADS products is under intensive investigation.

4.3 Resources

4.3.1 Alternative fuels

The transportation sector has an unprecedented task ahead in making effective and fast-acting carbon savings. Biofuels can play a role in the creation of a sustainable low-carbon transport system. The extent to which increased production of biodiesel and bioethanol will help to deliver viable, sustainable transport solutions will depend on how, and at what scale, it is implemented. Alternative fuels can be produced in several different ways. Today it is commonly agreed that only so-called "drop-in" fuels have a chance to replace petrochemical jet fuel within short term, i.e. those that can be used for current aircraft engines without modifications, and be blended with current jet fuel. For the implementation of non-drop-in fuels (e.g. liquefied gas) there are considerable challenges to overcome, such as impact on airports, airplanes and engines.

4.3.2 Hydrogen

The future of renewable hydrogen energy depends strongly on reduced costs for renewable energy production. Renewable hydrogen energy will enter the marketplace when and where it is cost effective compared to the other forms of energy.

5. Conclusions

To ensure growth of air transportation in an environmentally friendly way ("green" aircraft technology), EADS has to demonstrate being an eco-efficient industry, limiting the environmental impact to an acceptable level. This can only be achieved by introducing latest technology and innovation into EADS products to ensure a reduction of fuel consumption and therefore a reduction of emissions such as greenhouse gases, pollutants and noise.

References

IATA 2008 Report on Alternative Fuels.
IATA Technology Roadmap Report 3rd edition (June 2009).
ACARE: A Vision for 2020.
Air & Cosmos N° 1962, Dec. 2004.

BUILD-UP STRATEGIES FOR A HYDROGEN SUPPLY AND REFUELLING INFRASTRUCTURE INCLUDING A COMPARATIVE OUTLOOK ON BATTERY-ELECTRIC VEHICLES AND THEIR INFRASTRUCTURE REQUIREMENTS

R. Wurster, C. Stiller

Ludwig-Bölkow-Systemtechnik (LBST), Daimlerstrasse 15, 85521 Ottobrunn, Germany,
E-mail: reinhold.wurster@lbst.de

Abstract The paper outlines the resource situation and the long-term switch from a fuel to an electricity based energy system. Large volume storage of electricity over several days is identified as a key element for such an energy system. A comparison of sustainable road transportation concepts, mainly hydrogen fuel cell and battery-electric vehicles is presented. The hydrogen infrastructure build-up via modelling of the local hydrogen demand and allocation of refuelling stations, the optimisation of the H_2 supply infrastructure and an economic evaluation of the infrastructure results is described in more detail. The infrastructure requirements for battery-electric vehicles are comparatively assessed. Finally, an outlook on possible changes in customer attitude and external constraints possibly leading to more efficient and smaller vehicles is given.

1. Resource situation – the switch from a fuel to an electricity based energy system

The aggregate supply of fossil and nuclear energy is expected to peak by around 2015. After the peaking of the oil supply (today), the global supply of natural gas, coal and nuclear fuel is expected to reach a combined maximum by around 2020 – at the latest. (Zerta 2008) This will have significant impacts on the total energy supply. With peak oil we are entering into a transition phase towards a post-fossil energy area.

J. Wellnitz et al. (eds.), *Sustainable Automotive Technologies 2010*,
DOI 10.1007/978-3-642-10798-6_3, © Springer-Verlag Berlin Heidelberg 2010

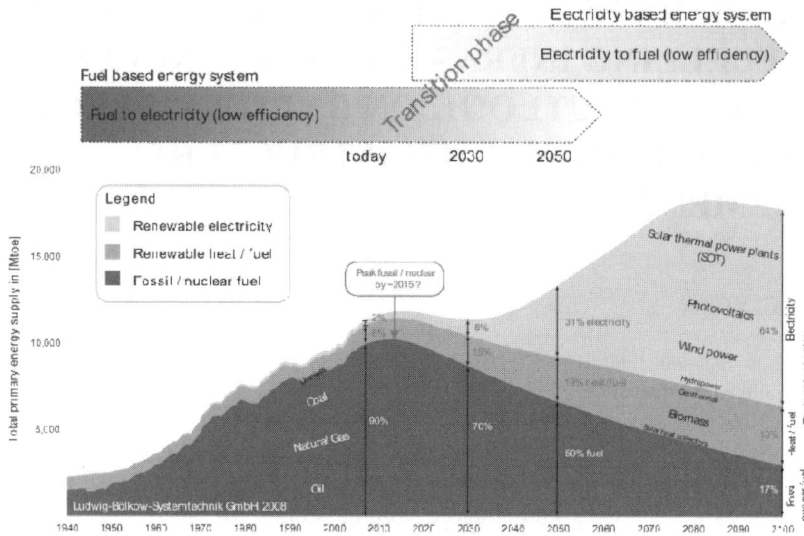

Fig. 1 Future primary energy supply (Schindler et al. 2008).

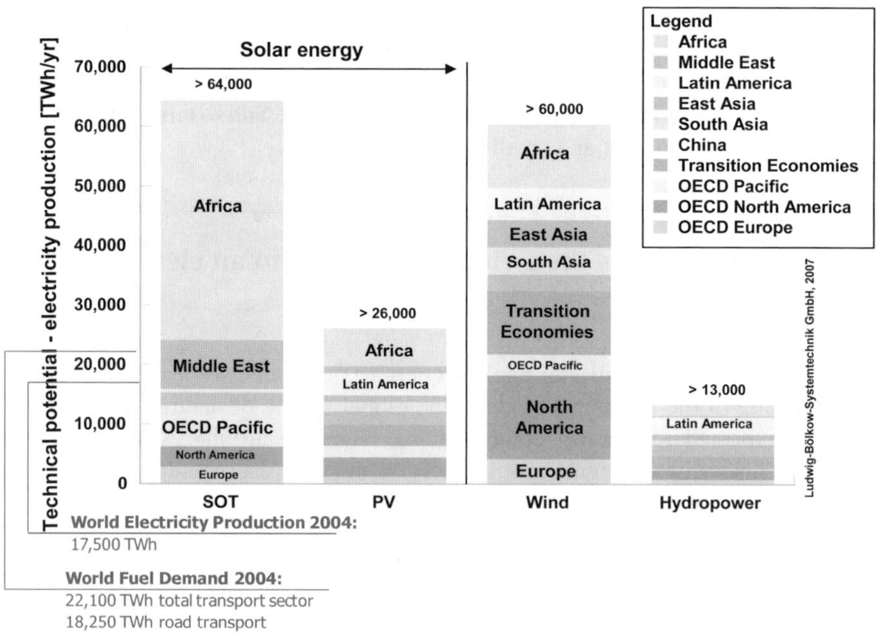

Fig. 2 Technical potential of renewable electricity – worldwide (Schindler et al. 2008).

However, in the long term, renewable energies will be able to provide more energy than all fossil and nuclear fuels will have been able to supply at their aggregate peak (see Figure 1 and Figure 2). The transition from fossil fuels to renewable energies is also a transition from primary energy fuels to electricity. Consequently, also the transport sector will be electrified to a large extent. Soon, road vehicles will start to switch to renewable electricity as a "primary energy source". Since a high share of fluctuating renewable electricity in the energy system is not manageable without a means of large-scale storage of electricity, hydrogen will become a very essential partner for the renewable energy world and especially for the future transport sector (Schindler 2008)

2. Large-scale storage of electricity

As Section 1 has shown, renewable electricity will become the major commodity in the future's energy system. Significant shares of this electricity will be of fluctuating nature requiring either storage, or support power capacity, otherwise optimum use and integration into the energy system will be difficult. Furthermore, also supply and demand will not match over the seasons and thus will require good management as well as some electricity storage. A recent study by the Power Engineering Society within VDE (Leonard 2008) has shown that the most cost effective electricity storage bridging disparities of in the order of one week or more will be hydrogen (see Figure 3).

The reference case 1 "*weekly storage*" (no more than one charge/discharge cycle per week) has shown that all storage technologies result in high electricity production costs. While pumped storage hydroelectric power stations would normally be the lowest cost option, in many countries there is practically no potential for expanding the necessary storage volume. Hydrogen storage in underground caverns (mostly salt caverns) is the feasible option at the lowest costs. In this case capacity-specific costs are lower because the energy storage density is higher by about one order of magnitude compared to adiabatic compressed air. Effective use of even limited cavern capacities would become feasible. Hydrogen cavern storage thus provides the possibility of a sustainable replacement for thermal power stations when it comes to the provision of reserve power. Hydrogen cavern storage furthermore can provide a source of cost effective renewable automotive fuel supply for FCVs.

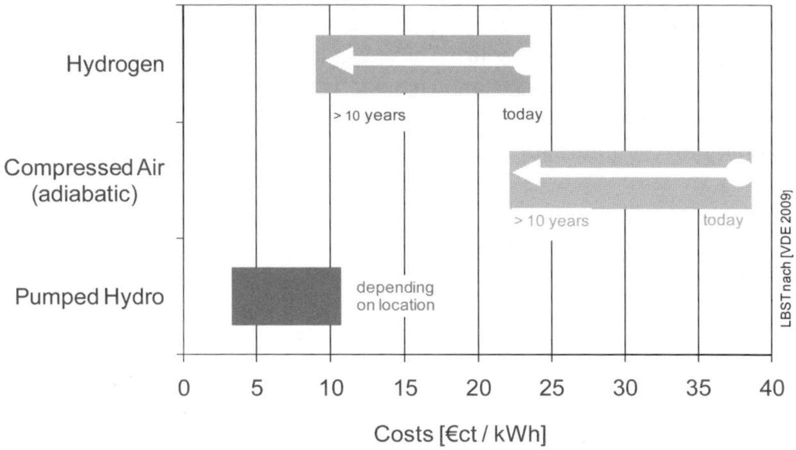

Fig. 3 Comparison of the full costs of storage systems for long-term electricity storage (Leonhard 2008).

3. Comparison of sustainable road transportation concepts

Generally, the best alternative transport fuel in terms of high conversion efficiency and low GHG emissions would be renewable electricity used in battery electric vehicles. Also with respect to yield per land area invested into fuel production, renewable electricity pathways, either for direct use in batteries or for electrolytic hydrogen production and use in fuel cell vehicles (FCVs), beat all biomass-derived fuel pathway and powertrain combinations by far (see Figure 4).

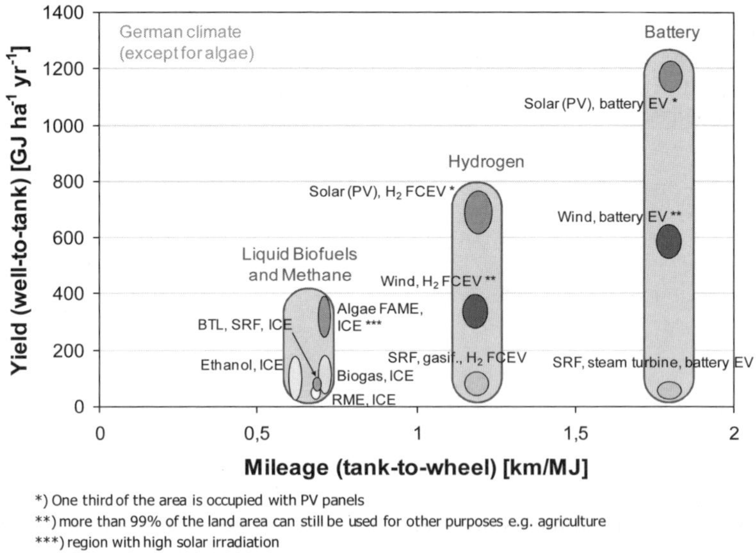

Fig. 4 Mapping of key performance criteria: "mileage" versus "yield" (LBST 2009).

But until now there is no battery technology for pure battery electric vehicles (BEVs) in sight which meets all requirements (long operating ranges, fast refuelling, low cost, zero emission, hot/cold climate operation, long lifetime). Therefore, electricity with an efficient battery storage may become an option for short to medium distance individual transport (150-300 km) in passenger cars if battery costs will drop from today's 600-800 €/kWh in small scale manufacturing to significantly below 300 €/kWh in mass manufacturing. Otherwise batteries would be prohibitively expensive especially for small city vehicles (as they would add easily 50% or more to the vehicle's base price) or could only be sold in luxury cars. Longer distance transport and propulsion of heavier vehicles might not become feasible merely with batteries due to weight and cost restrictions.

Here, hydrogen produced from renewable electricity (e.g. excess wind power converted to hydrogen and stored in underground salt caverns) probably will provide the best alternative transport fuel option in the future, in particular in combination with hybridised fuel cell electric vehicles (HFCVs) for medium to long distance travel in cars as well as for heavier road vehicles, like vans, buses and delivery trucks.

The energetic WtW efficiency of H_2-HFCVs is at least equal to that of best of class diesel vehicles today; already with a considerable GHG emission advantage (i.e. in the case of natural gas derived hydrogen about 30% lower GHG emissions can be achieved). On the other hand, the mere energetic WtW cycle efficiency of H_2-HFCVs is at least a factor of 1.5 lower than a BEV with the same operating range. But an important issue is sometimes overlooked or neglected: the much higher weight of the battery (see Figure 5) and the thereof resulting significantly higher specific propulsion energy requirements at comparably assumed autonomies.

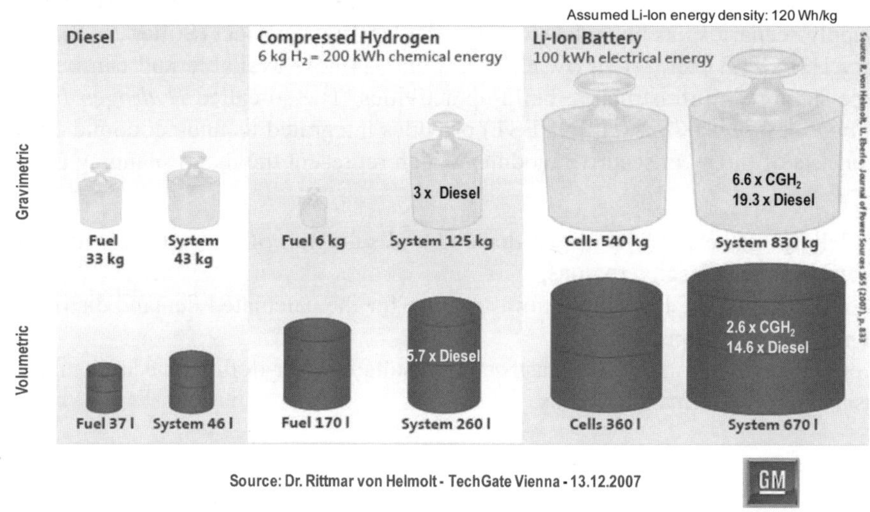

Fig. 5 Comparison of storage system densities (500 km).

If we assume a HFCV and a BEV, each with an operating range close to 500 km, then the primary energy requirements are about the same when natural gas is the primary energy source. If electricity is the energy "source" (e.g. electricity produced from wind power), then energy demand of the HFCV per km-driven is about 1.5 times higher than that of the BEV with the same driving range. If the battery-electric vehicle realistically is calculated at a much lower operating range of 150-300 km then its energetic advantage is gradually approaching a factor of 2.5, based on electricity as "energy source", and a factor of about 1.5 based on natural gas. But in the latter case (differing operating ranges) we are not talking about the same vehicle any more with regard to customer needs, because the assumed autonomy has been reduced to about half. Finally the customer has to decide which technology and thereto related characteristics are most acceptable.

In mass manufacturing both vehicle types (BEV and HFCV) might achieve comparable sales prices, with a slight tendency of BEV remaining somewhat more expensive at comparable autonomies.

4. Hydrogen infrastructure build-up

The European Commission co-funded project HyWays (www.hyways.de) and the German funded project GermanyHy (www.germanhy.de) have among others analyzed the build-up of a hydrogen vehicle market and the thereto related refuelling infrastructure and supply in the past years. They showed clearly that investment needs are not the show stopper for the hydrogen infrastructure. Nevertheless, a well coordinated and synchronized phase-in of large volumes of vehicles and refuelling stations is a prerequisite in order to avoid underutilized stations and economic losses.

A novel modelling approach leading to realistic regional hydrogen demand scenarios and supply scenarios has been developed in the NorWays project (Stiller 2008). The proprietary H2INVEST model (www.h2invest.com), is now available and can be used for analytic support in infrastructure build-up activities. The so-called *Hydrogen Infrastructure Venture Support Tool* (H2INVEST) provides integrated techno-economic evaluation and consists of three consecutive modules which represent the major planning and analysis steps:

1. Modelling of the local hydrogen demand and selection of concrete refuelling station locations in the selected regions,
2. Optimisation of the H2 supply infrastructure for the calculated demand distribution of the refuelling stations and
3. Economic evaluation of the infrastructure results for user-defined business units on the basis of common financial ratios.

4.1 Modelling of the local hydrogen demand and allocation of refuelling stations

In general, the modelling of demand for hydrogen as a transportation fuel is based on demographic development data as well as expected market penetration of the fuel cell vehicles in the relevant analysis areas. The single regions can be selected either manually in order to account for politically interesting but economically less attractive regions, or automatically by user-defined selection criteria for the build-up of the infrastructure. These selection criteria include a number of demographic factors such as population density, specific purchasing power as well as number of neighbouring regions with existing hydrogen supply. A single score is calculated in order to allow for a ranking of the regions. In this way H2INVEST can be used to simulate the introduction of the fuel cell vehicles and consequently the development of the specific hydrogen demand for the entire analysis timeframe within and between all areas.

This is followed by the selection of the corresponding refuelling station locations over the simulation time period. Based on the user-defined total number of refuelling stations in the analysis area the tool first calculates the minimum average driving distance to every hydrogen refuelling station in order to determine the respective number of refuelling sites in every region and time step. Then, using several assessment criteria such as available space, conventional fuel sales, visibility and accessibility of the site, conventional refuelling station locations are selected automatically for the upgrade with hydrogen dispensing equipment. Finally the actual demand for every hydrogen refuelling station is estimated analogically to the real consumption of the conventional fuels.

4.2 Optimisation of the H_2 supply infrastructure

The data on the hydrogen demand and refuelling stations are utilised in the second planning step in order to determine an optimum supply infrastructure. The essential optimisation criterion is represented by the minimisation of the actual total costs along the supply paths for all hydrogen refuelling stations. In this context the tool compares a number of H_2 production and distribution options for every refuelling station and automatically selects the best alternative in every time step. In this way it is possible to provide detailed information on the location, capacity and utilisation of each production facility and transportation unit for the entire analysis timeframe. Similarly, refuelling station sharp costs and analysis results can be provided.

In this context within the optimisation routine the tool takes into account not only different production technologies (e.g. electrolysis, steam reforming, gasification of biomass and utilisation of existing facilities generating hydrogen as a by-product) but also the fact that some technologies can be applied both at a centralised location or directly at the concerned refuelling station as onsite equipment. In case of the centralised supply the corresponding transportation of hydrogen can be accomplished either by specific trucks (GH_2/LH_2) or by a pipeline network. In addition, suitable onsite facilities can be combined with short pipelines in order to supply refuelling stations in the vicinity.

4.3 Economic evaluation of the infrastructure results

The results of the demand modelling and infrastructure optimisation can finally be used within the third analysis step for a detailed economic evaluation. In this context the tool allows for a flexible definition of different business units which may represent various investors along the entire hydrogen value chain. According to predefined settings related to the technology, value chain step and region the single production facilities, transportation units and refuelling stations can be assigned to business units. Then, based on specific transfer prices between the actors, the respective revenues and costs can be calculated as well as further financial key performance indicators for single investors. Moreover, the business analysis takes into account additional economic parameters such as target capital structure, working capital requirement and inflation. In this way the tool is able to provide detailed and realistic income statement, balanced sheet, statement of cash flows and break-even analysis for every involved investor.

First evaluations by industry also have shown that the build-up of a full coverage refuelling infrastructure with 1,000 hydrogen refuelling stations (HRS) in Germany would cost in the order of 1.5 – 2 billion Euro. Such an infrastructure can support about 1 million FCVs. The cost of one small HRS providing about 1,000 cars with fuel lies between 1.2 and 1.5 M€. For larger HRSs following larger vehicle populations and providing fuel to at least 5,000 cars cost reach about 3.5 M€ for each station. HRS typically refuel hydrogen vehicles in less than 3-5 minutes.

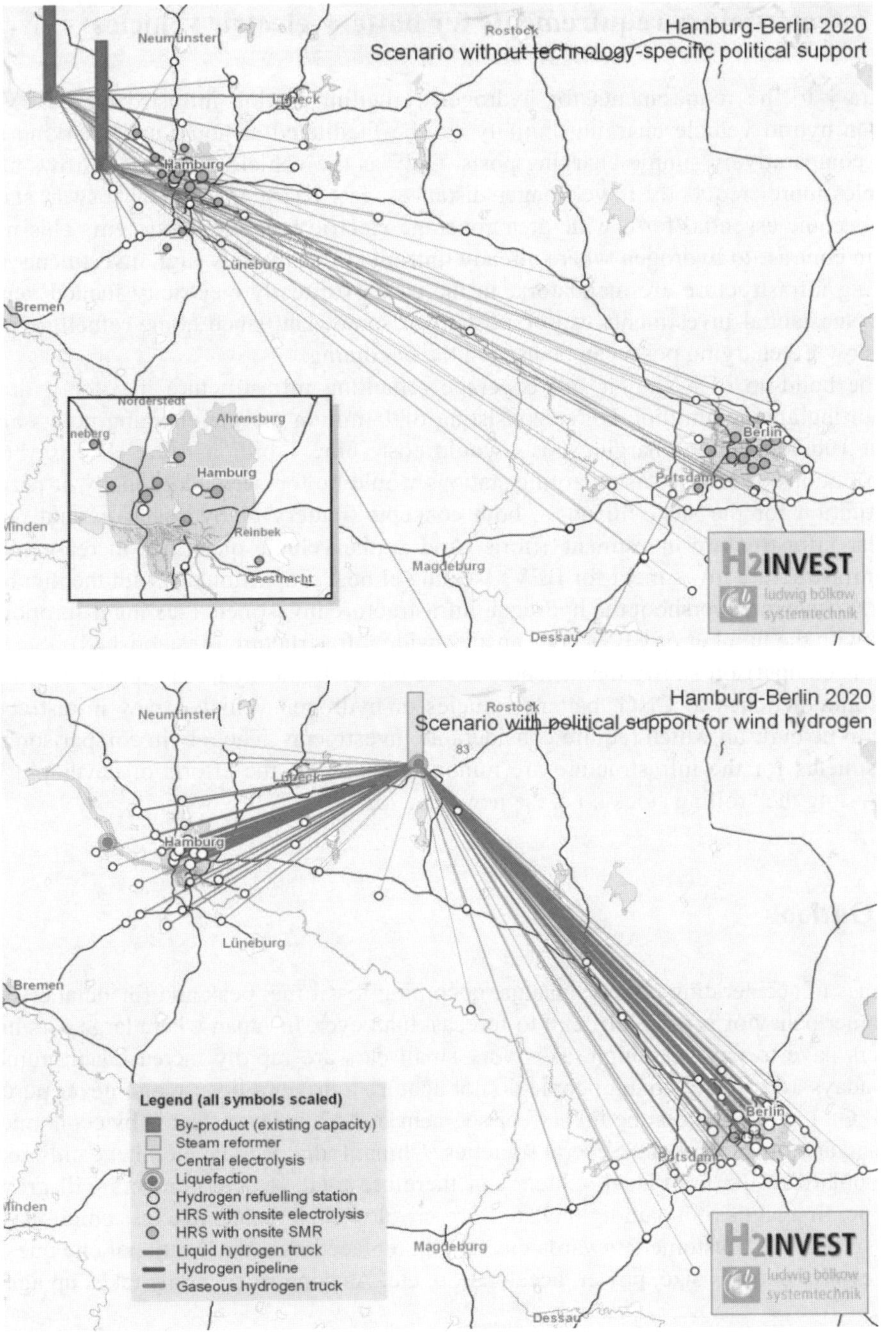

Fig. 6 Cost optimum hydrogen supply to Hamburg and Berlin with selective political support for wind-hydrogen (bottom) and without (top) (Stiller 2009).

5. Infrastructure requirements for battery-electric vehicles

Contrary to the requirements for hydrogen refuelling station infrastructure, BEV and plug-in hybrid vehicle charging initially can be facilitated at home or in parking areas with comparatively simple charging posts. Only as the vehicle population grows and as vehicles more frequently travel longer distances, fast charging or battery swap stations will become essential for a wide-area coverage electric fuel supply system. This means that in contrast to hydrogen where already upfront comparatively high investments in refuelling infrastructure are mandatory, in the case of directly electricity fuelled vehicles these substantial investments can be postponed somewhat, since home refuelling points and slow fill charging posts can be used in the beginning.

The build-up of a similar full coverage refuelling infrastructure like for hydrogen, supporting about 1 million BEVs consisting of 1 million private charging points and of about 100,000 public charging posts would cost some 1 billion Euro[1]. In case battery swap stations or battery fast charging stations would be foreseen additionally or partly as substitution for the slow fill posts, both concepts (battery or hydrogen) would require similar infrastructure investment efforts. And as the vehicle quantities increase further, the infrastructure investment for BEVs (scaling almost proportionally with the number of BEVs) can even overshoot the hydrogen infrastructure investment (scaling disproportionately with the number of BEVs once an area-wide infrastructure is established).

This comparison shows that as soon as one leaves liquid hydrocarbons as automotive fuels and switches to CNG, battery vehicles or hydrogen vehicles, new infrastructures have to be built up which require considerable investments. Though, in comparison these investments for the infrastructure are minor compared to the efforts of developing and purchasing the "rolling stock", i.e. the new vehicles.

6. Outlook

In times of accelerating climate change, upcoming fossil fuel peak and financial crisis the customer behavior is more difficult to forecast than ever. In Japan where large cars in city centers have to pay very high taxes very small cars are rapidly increasing in numbers. Nowadays even many young people do not acquire a driver's license and never purchase a car. On the other hand, the luxury car segment in Asia is less affected by economic crisis than any other segment. Several societies (China, India, Russia, etc.) have only recently popularized the individual vehicle and therefore total vehicle numbers will grow. In Europe, first signs of a paradigm change are visible which results in a less emotional attitude of potential customers towards the vehicle replaced by a more rational one questioning features such as size, power, acceleration, etc. As soon as economy takes up again, it

---

[1] Slow fill charging posts supporting between 4 and 5 cars cost in the order of 4-5,000 € each. Fast fill charging posts (15 kWh in 15 minutes) supporting about 50 cars will reach 25-35,000 € each depending on manufacturing volumes.

will become visible that oil production capacities have not been extended over the last years and that growing demand on a global scale will be less likely to be covered. This and the implications of climate change will force us towards more efficient vehicles and then to new energy sources for new fuels – at the long run electricity becoming the base-commodity.

References

Helmolt R. v. and Eberle U. (2007), Fuel Cell Vehicles: Fundamentals, System Efficiencies, Technology Development, and Demonstration Projects, Hydrogen Technology - Mobile and Portable Applications, Series: Green Energy and Technology, Aline Léon (Ed.), ISBN: 978-3-540-79027-3, pp. 273-290, doi: 10.1007/978-3-540-69925-5_9

Jackson N. (2009) Innovation in Sustainable Transport, European Future Energy Forum, Bilbao, 9-11 June 2009

Leonhard W. et al. (2008), Energy storage in power supply systems with high share of renewable energy sources – Significance, state of the art, need for action, ETG Energy Storage Task Force, VDE Association for Electrical, Electronic & Information Technologies, Frankfurt, December 2008

Simpson A. (2009) Electric Vehicles and their Renewable Connection - How Australia Can Take Part in the Green Revolution, RMIT Seminar – 22 June 2009

Thomas CE. (2009) Fuel cell and battery electric vehicles compared, International Journal of Hydrogen Energy (2009), doi:10.1016/j.ijhydene.2009.06.003

Schindler J. et. al. (2008) Hydrogen and Fuel Cells as strong Partners of Renewable Energy Systems, European Hydrogen Association (EEHA) and Germany Hydrogen and Fuel Cell Association DWV)

Stiller C. (2009), Modelling hydrogen infrastructure build-up - LBST develops H2INVEST – Comprehensive tool for planning and analysis of the infrastructure for hydrogen refuelling stations, HyWeb, 2009-05-20 (www.hyweb.de)

Zerta M., Schmidt P., Stiller C., Landinger H. (2008) Alternative World Energy Outlook (AWEO) and the role of hydrogen in a changing energy landscape, International Journal of Hydrogen Energy, 33/12, 3021-3025.

Stiller C., Bünger U., Møller-Holst S, Svensson AM., Espegren KA, Nowak M. (2008) Pathways to a hydrogen fuel infrastructure build-up in Norway. International Journal of Hydrogen Energy, in press.

Wurster R., Stiller C. and Bünger B. (2008), Wasserstoffinfrastruktur für Straßenfahrzeuge - Neuer Ansatz zur regionalen Modellierung von Nachfrage- und Versorgungsszenarien, Energy 2.0-Kompendium 2009 |www.energy20.net

APPROACH FOR THE SOLUTION OF A SIMPLIFIED REISSNER THEORY FOR AUTOMOTIVE APPLICATION

S. C. Bruckmeier[1], J. Wellnitz[1,2]

[1]Privat-Research-Institute for Technology and Artistic Design e. V.
Marie Curie Straße 6, 85055 Ingolstadt, Germany
stephanie.bruckmeier@itd-in.de, www.itd-in.de

[2]University of Applied Sciences Ingolstadt
Esplanade 10, 85049 Ingolstadt, Germany
joerg.wellnitz@fh-ingolstadt.de, www.fh-ingolstadt.de

Abstract To calculate the behavior for one of the most promising new sustainable and recyclable materials in the automotive industry the elastic plate Theory of Reissner is applied. The so-called CMA (Core Metal Alloy) composite combines lightweight advantages with wide availability. Since the theory of Reissner results in a difficult system of differential equations a slight modification is introduced. The boundary value problem is solved under static conditions and specified given load. The solution consists of well-known analytical functions and provides sufficient constants for possible boundary adjustments.

1. Introduction

The main driving forces for the development of a new material or composites are not only the lightweight aspects but also the availability, being a major determining factor for its success in the automotive industry. Having this in mind the idea to create a new laminate, that combines thins layers of stone and metal, the so-called CMA, arose. Stone as a lightweight structural element is almost unlimited in availability, since the earth is the largest rocky body of the four solar terrestrial planets, both in terms of size and total mass.

The new sustainable and recyclable material is shown in Figure 1. It shows huge potential with application in aerospace, automotive, and civil engineering (see Figure 2 for example).

J. Wellnitz et al. (eds.), *Sustainable Automotive Technologies 2010*,
DOI 10.1007/978-3-642-10798-6_4, © Springer-Verlag Berlin Heidelberg 2010

Fig. 1 CMA-Composites.

Fig. 2 CMA-Application as heat shields in turbines.

Considering CMA as a common structural element it is necessary that the theoretical behavior, in particular the kinematics of its deformations be described by using adequate theories. While the classical plate theory based on the Kirchhoff hypothesis neglects transverse shear strains, the theory from Reissner considers the problem of shear deformation in plates and additionally includes normal transverse stress effects. Thus the Reissner theory is more accurate for thicker plates or materials with a relatively soft interlaminar shear modulus as in the CMA-Composite.

2. The Reissner theory

We assume an isotropic, rectangular plate with a uniform thickness *h*, which is very small compared to the length *a* and width *b* ($h << a, b$). Further, the total deflections of the plate are assumed to be small in comparison with its thickness *h*. Consequently the first order theory can be applied.

Analogous to the classical plate theory, linear stress is assumed by Reissner through the plate thickness as follows:

$$\sigma_{\alpha\beta} = 12z/h^3 M_{\alpha\beta} \tag{1}$$

With consideration of the Cauchy equilibrium equations a square polynomial represents the shear stress in respect of *z* direction:

$$\sigma_{\alpha 3} = \frac{3Q_\alpha}{2h}\left[1 - 4\left(\frac{z}{h}\right)^2\right] \tag{2}$$

and the normal stress is then given by:

$$\sigma_{33} = \frac{3p}{4}\left[-\frac{2}{3} + \frac{2z}{h} - \frac{8}{3}\left(\frac{z}{h}\right)^3\right] \tag{3}$$

where p is the applied distributed transverse load. Introducing the three variables w (deflections of plate) and the rotations ψ_x, ψ_y, which are determined by applying the virtual work principal yields:

$$\psi_\alpha = \frac{1}{M_\alpha}\int_h \sigma_\alpha U_\alpha dz \quad \text{and} \quad Q_\alpha w(x,y) = \int_h \sigma_{\alpha z} W(x,y,z)\, dz. \tag{4}$$

Hereby the stress that causes real displacement through the thickness produces the same amount of energy equal to an average, fictitious variable e.g. the rotation ψ_α corresponding to its internal force. Replacing the equations in (4) by (1) and (2) and the relations:

$$\psi_\alpha = \frac{12}{h^3}\int_h zU_\alpha dz \qquad \text{and} \qquad w = \frac{3}{2h}\int_h [1-(\frac{z}{h/2})^2]W dz\,, \tag{5}$$

are obtained.

Additionally with the aid of the constitutive equations for linear elastic material a system of three differential equations is found, by conducting various integrations with respect to z and finally taking the dynamic equations (of motion) into account. The three introduced principal unknowns of the dynamic plate problem are determined by:

- The first Reissner differential equation (4th order):

$$\mathrm{D}\Delta\Delta\mathrm{w} - \frac{3-\mathrm{v}}{1-\mathrm{v}}\frac{\rho\mathrm{h}^3}{12}\Delta\frac{\partial^2}{\partial t^2}\mathrm{w} + \frac{\rho^2\mathrm{h}^3}{12\mathrm{G}}\frac{\partial^4}{\partial t^4}\mathrm{w} + \frac{5}{6}\rho\mathrm{h}\frac{\partial^2}{\partial t^2}\mathrm{w} = \mathrm{p} - \frac{2-\mathrm{v}}{1-\mathrm{v}}\frac{\mathrm{h}^2}{10}\Delta\mathrm{p} + \frac{\rho\mathrm{h}^2}{10\mathrm{G}}\frac{\partial^2}{\partial t^2}\mathrm{p} \tag{6}$$

- The second and third Reissner differential equation (2nd order):

$$\psi_\alpha + \frac{(1+v)}{(1-v)}\frac{\partial}{\partial\alpha}\frac{\partial}{\partial\lambda}\psi_\lambda - \frac{10}{h^2}\left(\frac{\partial w}{\partial\lambda} + \psi_\alpha\right) + \frac{12v}{5Eh}\frac{(1+v)}{(1-v)}\frac{\partial p}{\partial\alpha} = \frac{\rho}{G}\frac{\partial^2\psi_\alpha}{\partial t^2} \tag{7}$$

These equations represent inhomogeneous, linear, partial differential equation of 6th order with constant coefficients, which also allows the satisfaction of three independent boundary conditions at each edge of the elastic plate.

3. Approach for an analytical solution

We construct the analytical solutions in known elementary functions that exactly satisfy simplified Reissner differential equations for the static plate problem. Once the solution had been found the internal forces and moments could be calculated.

3.1 The Modification of the governing Reissner equations

Because of the complexity of the system of differential equations shown in (6) and (7), even for static problems, the derivate solutions in this present study are based on slight modifications to the equations, according to the Gauss curvature. The idea is to uncouple the second and third Reissner differential equations, which is given by a specific derivation term as shown in following figure and terms:

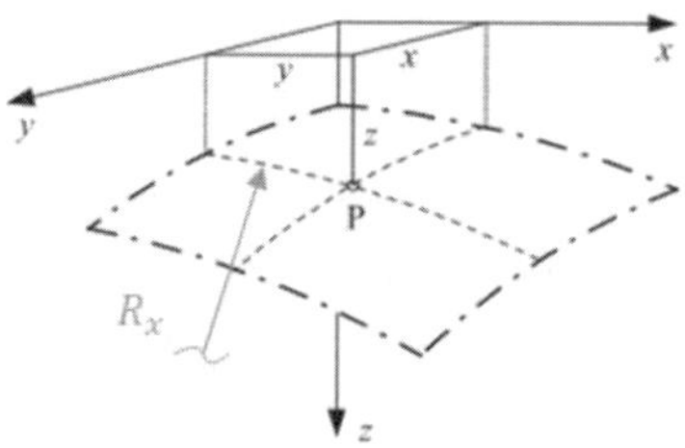

$$\kappa_\alpha = \frac{1}{R_\alpha} << 1: \quad \Rightarrow \quad \frac{\partial}{\partial x}\frac{\partial \psi_y}{\partial y} \text{ bzw. } \frac{\partial}{\partial x}\frac{\partial \psi_x}{\partial y} \approx \frac{1}{1000} \ldots \frac{1}{100} \tag{8}$$

Fig. 3 Assuming a very small imperfection of an elastic plate.

This derivation term in equation (8) is replaced by a newly introduced parameter *P*. To do this we suppose that the rectangle plate has a very small pre-curvature and thus the Gauss curvatures are very small. This approximated value (i.e. the constant parameter *P*) is orientated on a given small pre-curvature plate. Moreover it is assumed that the imperfection is so small that the theory of first order can still be applied.

3.2 Specified external load

The boundary value problem was being solved for polynomials distributed loads:

$$\bar{p}(x,y) = \sum_{k=0} p_k\, x^k + \sum_{i=0} r_i\, y^i \tag{9}$$

Various shapes can be created and an example is demonstrated in Figure 4:

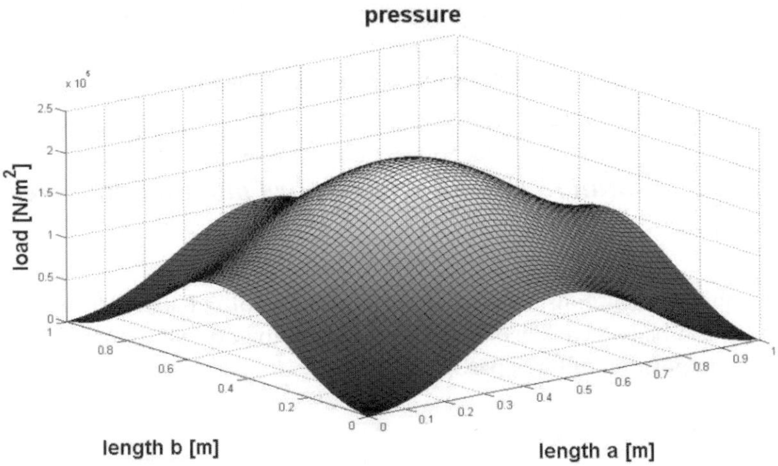

Fig. 4 Specification of the external load $\bar{p}$.

This certain load approach represents a common load type in the automotive industry e.g. ballast. In this study a peak pressure of 200,000 $\frac{\mathrm{N}}{\mathrm{m}^2}$ is applied in the middle of the plate.

3.3 Displacement function

The solution for the first Reissner equation consists of two parts, which are super positioned:

$$w = w_h + w_p \tag{10}$$

The deflection w_h must satisfy the homogeneous differential equation in (6). This solution had already been found, because in case of static conditions it is identical to the Kirchhoff plate problem. As provided in Girkmann (1963) the solutions consist of well known superposition's of hyperbolic and trigonometric terms.

The second (particular) solution w_P is introduced to satisfy the applied power series load $\bar{p}$. Hence the partial solution is also developed as a power series approach as follows:

$$w_p = B_0 + B_1 x + B_2 x^2 + B_3 x^3 + \sum_{k=0} B_{k+4}\, x^{k+4} + \tag{11}$$

$$+ C_0 + C_1 y + C_2 y^2 + C_3 y^3 + \sum_{i=0} C_{i+4} y^{i+4}$$

The coefficient B_{k+4} and C_{i+4} are specified by a simple comparison of the coefficients from the external load.

Both the homogeneous solution and the partial solutions provide a sufficient amount of coefficients (even more than is necessary in terms of the order of the integration problems) which allow adjustments on the boundaries.

3.4 Analytical solution for second and third modified Reissner equation

With the obtained solution w the second and third modified static REISSNER equations in (7) can be solved. The solution for the rotation normal to the middle plane is also a superposition of three partial solutions:

$$\psi_\alpha = \psi_{\alpha h} + \psi_{\alpha w_h} + \psi_{\alpha p w_p} \tag{12}$$

While the $\psi_{\alpha h}$ are general solutions for the homogenous second and third modified equations the other partial solutions satisfy the inhomogeneous part that derivates from the deflection function (more precisely $\psi_{\alpha w_h}$ is for the homogenous partial solution and $\psi_{\alpha p w_p}$ for the particulate partial function). The function $\psi_{\alpha h}$ provides constants which should also be determined from the boundary conditions and selected as good as possible.

4. Application for the CMA

The CMA-Plate is composed of multiple layers of aluminium and stone. To provide a homogeneous material for the theory the simple mixture rules that are commonly used in micro mechanics are applied. Applying a spring model determines a practical approximation of material values. An example for the Young's Modulus $\breve{E}$ is shown in equation (13), which is determined by the thickness of the individual layer.

$$\breve{E} = E_{stone} \frac{t_{stone}}{t} + E_{Alu} \frac{1-t_{stone}}{t} . \tag{13}$$

Because of the layer composition, the CMA results in orthotropic material properties. In fact no preferred direction exists among the material in its plane, except in the transverse direction. This effects the linear constitutive equations. This is neglected in the present study and Hooke's law for isotropic material is assumed.

5. Conclusion

The governing equations of the Reissner theory for rectangular, homogenous and isotropic plates have been applied to describe the material behavior of new composite CMA. For the solution an approximation has been introduced.

Further works contain the investigation of sensitivity of the introduced parameter P and its effect on the results as an appropriate approximation.

In addition to the static solution approaches for the dynamic problem have also been worked on.

Note that the derivate analytical solution provides constants of integration that will be evaluated using the boundary conditions or regarding transient problems initial conditions are also necessary.

The composite CMA as a lightweight structure element is currently used for Honda bonnets and heat shields.

References

Reissner E. (1945) The Effects of transverse shear deformation on the bending of elastic plates. Journal Appl. Mech. 12.

Panc V. (1959) Theories of elastic plates. Noordhoff, Leyden, The Netherlands

Girkmann K.(1963) Flächentragwerke. 6. Auflage, Springer, Wien.

Emmerling A. (1989) Technische Mechanik und Flächentragwerke. Vorlesungsskript, Fakultät für Luft- und Raumfahrttechnik, Institut für Mechanik, Universität der Bundeswehr München, Vorlesungsskript.

Walthelm U. (1978) Lösung der REISSNER`schen Plattentheorie für isotrope rechteckige Platten mit Verwendung von reellen Eigenwerten, sowie Vergleich mit der klassischen Plattentheorie und dem Balken. Ph.D. Diss., Fachbereich Bauingenieur- und Vermessungswesen der technischen Universität München.

Wellnitz J. persönliche Mitteilung, 01.06.2009.

Bruckmeier S. (2009) Ein Lösungsbeitrag zur Plattentheorie nach Reissner. Master Thesis, Hochschule Ingolstadt, Ingolstadt, Deutschland.

GREENER ROADS BY TALKING TRAFFIC LIGHTS: KNOWLEDGE ABOUT QUEUE LENGTH AND UPCOMING TRAFFIC LIGHT SIGNAL

C. Dittrich[1], J. Lüßmann[1], F. Busch[1], R. Mänz[2], R. Stahlmann[3], C. Menig[3]

[1]Chair of Traffic Engineering and Control, Technische Universität München,
Arcisstraße 21, D-80333 München, Germany,
E-Mail: claudia.dittrich@vt.bv.tum.de, jonas.luessmann@vt.bv.tum.de, fritz.busch@tum.de

[2]Department of Computer Science 7, Friedrich-Alexander-Universität Erlangen-Nürnberg,
Martensstraße 3, D-91058 Erlangen, Germany,
E-Mail: robert.maenz@ini.fau.de

[3]AUDI AG, I/EB-11, D-85045 Ingolstadt, Germany,
E-Mail: rainer.stahlmann@audi.de, cornelius.menig@audi.de

Abstract This paper focuses on two applications; urban traffic state estimation and green light optimal speed advisory, which have been implemented in the city of Ingolstadt, Germany. These Car2X applications use communication between vehicles and the infrastructure to inform the driver about the current traffic state within the urban road network and the optimal speed when approaching signalized intersections. Previously, a green light optimal speed advisory was implemented in the Audi project TRAVOLUTION. As part of the continuation of the project some new aspects have been implemented to improve the speed advisory, all focusing on the common goals of pollution and travel time reduction. Two of these new aspects are introduced here. The first is the signal change prediction and the other is queue length estimation.

1. Introduction to TRAVOLUTION

Moving through urban areas can be exasperating. Imagine there are several routes possible, but you took the one jammed with traffic, because you didn't know. And right at the moment you stop at a red traffic light it changes to green. TRAVOLUTION is an ongoing Audi project tackling scenarios where urban mobility can be improved. The pilot region is Ingolstadt in Germany – the home town of Audi. Traffic lights and vehicles are equipped there with a system enabling vehicle to infrastructure communication. There are two main applications. Urban traffic state estimation collects information throughout the city and classifies the traffic state in categories like free flow and jammed. Green light optimal speed advisory informs the driver about the optimal speed to pass the next traffic light

J. Wellnitz et al. (eds.), *Sustainable Automotive Technologies 2010*,
DOI 10.1007/978-3-642-10798-6_5, © Springer-Verlag Berlin Heidelberg 2010

without stopping. Not only does this smooth traffic flow, it also saves fuel, cutting urban vehicle emissions. This paper concentrates on two aspects that improve the speed advisory. Signal change prediction is necessary in order to give advice to the driver even at adaptive traffic lights. Without having a queue length estimation, the use case fails in dense traffic situations. As the queue length is also an important input to traffic state estimation, the applications are connected – like talking traffic lights communicating to vehicles.

2. Queue length estimation

Knowledge about queue length is an important aspect within the optimal speed advisory, as the algorithm has to consider the delay resulting from vehicles queued at an intersection. Therefore we are currently developing a method for queue length estimation.

2.1 State of the art

Work is still ongoing to optimise the recognition and estimation of queue lengths.

Priemer and Friedrich developed a method for online queue length estimation based on probe vehicle locations of communicating cars in the queue (2008). Additional data comes from the traffic lights (time of signal change), the loop detectors (time of continuous occupancy of the detector) and from statistics (reaction time and acceleration ability).

Comert and Cetin (2009) also approach queue length estimation by using probe vehicle location. They estimate the queue length based on the penetration of probe vehicles in the urban street network, indirectly drawing on vehicle distributions. Considering different penetration rates and the position of the last vehicle in line, they deduce the number of vehicles queued upstream of the probe vehicle.

Mück has invented a patented method (Mück 2007) for determining queue length. In contrast to the method of Comert and Cetin or Priemer and Friedrich, Mück's algorithm is based on detector data, especially the filling time. Mück found out a linear relationship between the queue length and the occupancy value of the loop detector. The filling time is the elapsed time between the beginning of red time and continuous occupancy of the detector. By comparing the actual filling time with a reference filling time, he derives the queue length. (Mück 2002)

2.2. Simulation

To prove the benefit from queue length information for the green light optimal speed advisory system, we simulate traffic flow on two 240 m distant consecutive intersections with fixed time signal control and low traffic load in Bremerhaven. The formation of a queue at the intersection where a speed advisory will be given is shown in Figure 1.

Fig. 1 Intersection under consideration.

The coordinated signal plans and the queue length characteristic are represented in Figure 2. The coordination for this direction is not optimized on green time end, because of the green wave in the opposite direction.

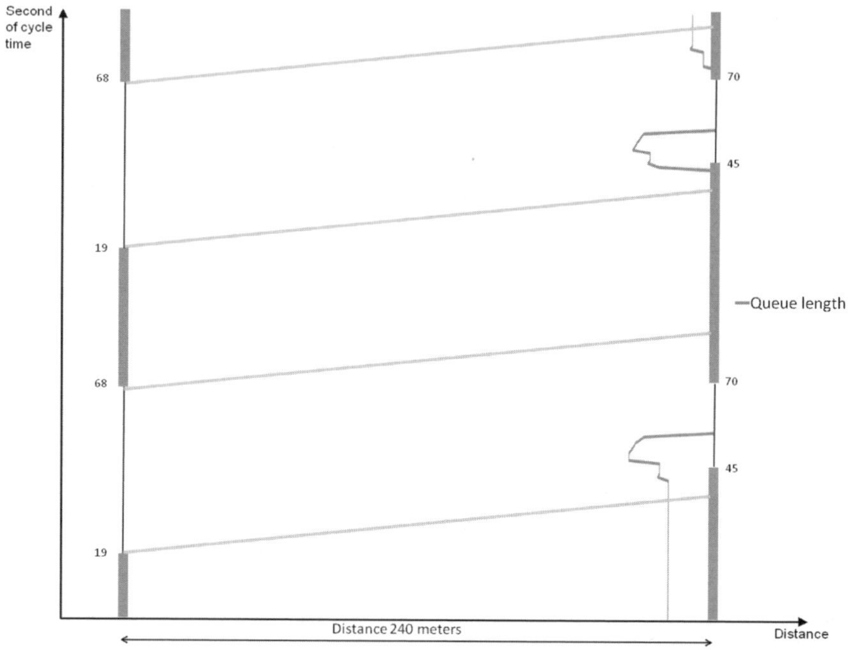

Fig. 2 Coordinated signal plans and queue length characteristic.

The graph shows the queue length in meters measured every second. Maximum queue length in the simulation is 40 meters. In the following it is assumed that we know the exact queue length 17 seconds in advance, which is the time a vehicle needs to get to the next intersection at maximum permitted speed of 50 km/h.

Based on the characteristic of the queue length, a different speed advisory is given to approaching vehicles at the stop line of the upstream intersection for different points in time.

2.3 Simulation results

The graphs in Figure 3 show the impact of the queue length on the speed advisory. The indicated lines restrict the speed to a lower and an upper limit, assuming a reasonable speed advisory within a range of 30 km/h to 50 km/h.

During the time a queue is present, the maximum speed that can be driven to reach the following intersection during green time differs noticeably from the speed advisory without queue length consideration. The minimum speed, which can be advised, does not differ as the queue did not last until the end of the green time. Figure 4 shows the difference of maximum speed advisory with and without considering the queue length. It can be seen that the difference on this short distance with relative short tailback already reaches 6 km/h. It is therefore important to include the queue length in the speed advisory.

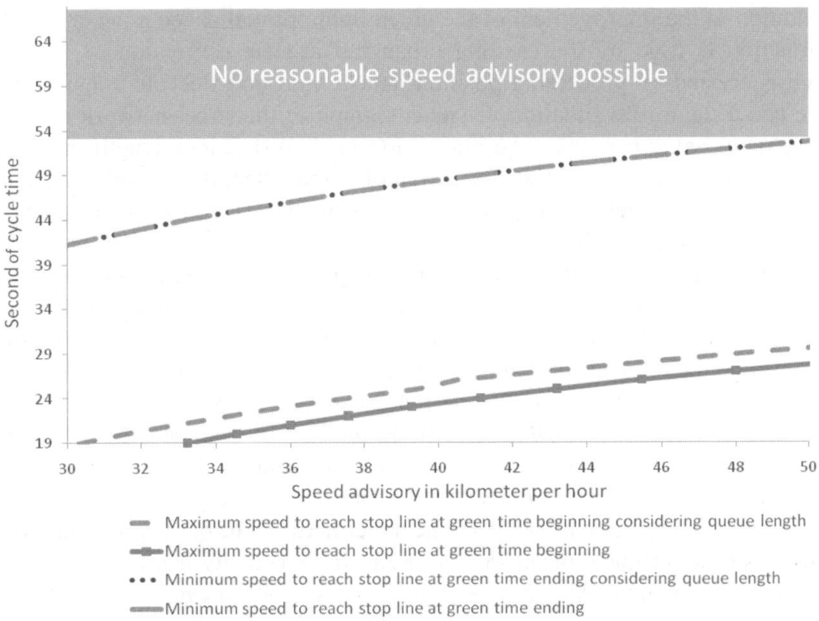

Fig. 3 Range of speed advisories.

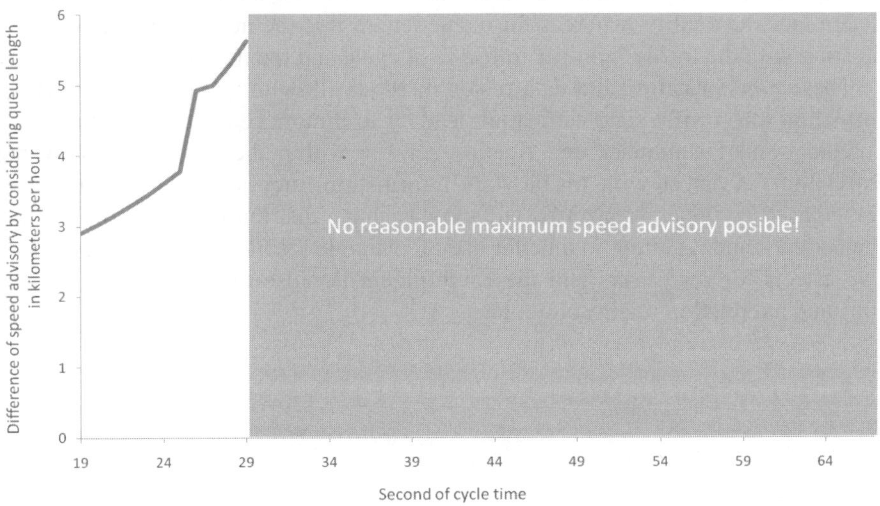

Fig. 4 Difference of speed advisory.

2.4 Integration of tailback information into the urban traffic state estimation

In addition to the improvement of the green light optimal speed advisory, tailback information is used by the existing urban traffic state estimation of Ingolstadt, which is created by processing position data from a taxi fleet, to enhance knowledge about the traffic situation. It yields a graph of the street network containing one velocity value per link. The integration method of queue length information consists of two main steps: transformation of tailback length into velocity, and data fusion of velocities from queue length estimation and taxi floating car data (FCD) on each network branch.

For transformation of the time-continuous queue length estimation into velocity, the mean waiting time per cycle is determined. Subsequently, the mean waiting time plus deceleration time, acceleration time and time of driving (time needed to pass through the route section in front of the traffic light with maximum speed) are used to derive the travel time of one representative car on the link in front of a signalized intersection. Finally, velocity on the link in front of the traffic light is calculated from travel time and length of the road section.

Estimation of an updated traffic state is performed every five minutes, requiring one velocity value per branch of the street network to be determined. For this purpose, a time weighted mean value of each single velocity from queue length estimation is calculated. All taxi speed values form one mean value. If there are floating car data and tailback length available on the same link, data fusion is applied. This is done by weighting mean values of both floating car data and the transformed tailback information. The weight can be thought of as a kind of quality that is calculated for each measurement. For FCD the number of measurements determines the quality, whereas for queue length it is the number of cars communicating with the traffic light per traffic light cycle and the number of lanes.

These transformation and data fusion methods allow integration of tailback information into traffic state estimation leading to a more reliable result due to the combination of stationary and floating sensors within the urban road network. Providing the driver with traffic state information, they are able to select their route based on the current traffic situation, easing a distribution of the load within the network and resulting in a better use of available resources. This leads to positive effects for road users and the environment because of a reduction of travel time and a reduction of air pollution.

3. Signal change prediction

3.1 State of the art

Former signal change prediction used in the TRAVOLUTION optimal speed advisory simply took into consideration the exported values of the traffic light controller. By getting these values the roadside-unit was able to follow the current traffic light controller behaviour. However, this information is not enough to predict the imminent behaviour. By merging the data coming from the traffic light controller with the knowledge of the chronological sequence of the signal changes, a prediction of the next 5 to 15 seconds was possible in case the signal change process had already started. The rest of the prediction was done by using the controller behaviour of the last program sequence. This prediction method works well if the traffic state is changing slowly and no abrupt interactions are arising that have not taken place in the last cycle, such as prioritization of public transports.

3.2 Enhanced signal change prediction

To improve the reliability of the prediction, the prognosis applies a stochastic method. This enables the optimal speed advisory to process inaccuracies resulting from the traffic adaptation of the signal control. Therefore more detailed information will be transmitted to the vehicles.

To achieve the signal change prediction, the algorithm will use much more information like historical information representing the behaviour of the traffic light controller to events like approaching busses. Additionally, traffic load curves will be used to enable the algorithm to deal with upcoming periods like rush hour or lower weekend traffic.

The result will not be a stringent prediction of the traffic light controller behaviour, but rather a representation of the current knowledge of the upcoming behaviour. Therefore likelihoods are used to consider inaccuracies in the prediction and communicate them to the vehicle where they can be dealt with correctly.

As can be seen in Figure 5 the quality of the prediction algorithm is represented by the slope within the transition region (time from 10s to 20s).

An optimal prediction is marking the exact time for the upcoming signal change with a Heaviside step function.

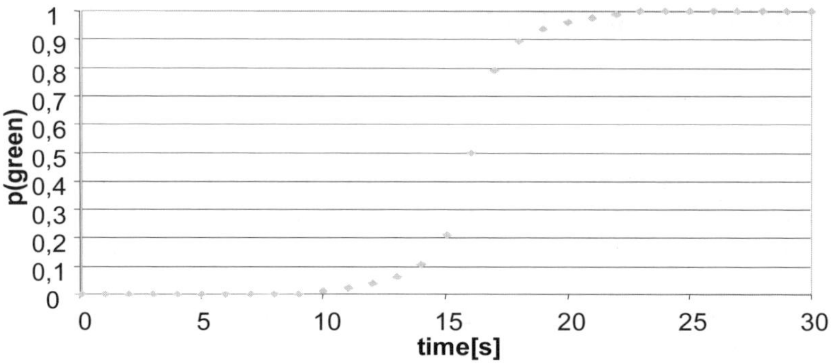

Fig. 5 30 sec. prediction with likelihood for green (Menig *et al.*, 2008).

4. Conclusion

In this paper a microscopic simulation is presented that demonstrates the benefit of tailback information on the green light optimal speed advisory.

Furthermore, tailback information is used to improve the already existing urban traffic state estimation of Ingolstadt. Transformations of tailback information into velocity and data fusion have been described.

In order to improve the reliability of the green time prediction within the green light optimal speed advisory, a stochastic method is needed. Ideas to this effect have been presented in the paper.

Overall, the enhancement in the optimal speed advisory improves the reduction of environmental pollution and inner city traffic jams, by minimizing the number of unnecessary stops at signalized intersections.

References

Comert G. and Cetin M. (2009) Queue length estimation from probe vehicle location and the impacts of sample size. *European Journal of Operational Research*, 197(1), p. 196–202.

Mück J. (2002) Schätzverfahren für den Verkehrszustand an Lichtsignalanlagen unter Verwendung halteliniennaher Detektoren. *Straßenverkehrstechnik*, 11/02, p. 613-618.

Mück J. (2007) Method for determining a queue identification number and for determining the length of the queue. United States Patent US7263435, published 28 Aug.

Priemer C. and Friedrich B. (2008) Optimization of Queuing Models within Adaptive Signal Control via C2I-data. In *Proceedings of the 10th International Conference on Applications of Advanced Technologies in Transportation*, 27-31 May, Athens.

Menig C., Hildebrandt R. and Braun R. (2008) Der informierte Fahrer – Optimierung des Verkehrsablaufs durch LSA-Fahrzeug-Kommunikation. Forschungsgesellschaft für Straßen- und Verkehrswesen (Hrsg.) In *HEUREKA '08, Optimierung in Verkehr und Transport - Tagungsband*, FGSV Verlag, Cologne.

SUSTAINABLE ANTENNAS – A WIDEBAND ANTENNA FOR VEHICULAR APPLICATIONS

B. D. Pell[1], E. Sulic[2], W. S. T. Rowe[1], K. Ghorbani[1], S. John[2], R. Gupta[3], K. Zhang[4]

[1]School of Electrical & Computer Engineering, RMIT University
[2]School of Aerospace, Mechanical & Manufacturing Engineering, RMIT University
[3]School of Civil & Chemical Engineering, RMIT University
[4]School of Mathematical & Geospatial Sciences, RMIT University
GPO Box 2476, Melbourne, VIC 3001, Australia

Abstract Use of wireless systems in vehicles has significantly increased in the past decade and by all appearances this trend will continue into the foreseeable future. Several decades ago, a simple protruding "mast" type antenna was the only antenna required in a vehicle, and was used for the purposes of receiving analog radio. Recent years have seen new types of antennas fitted to vehicles to accommodate new wireless services. We discuss a recently published antenna with extremely wide impedance bandwidth, the Planar Inverted Cone Antenna (PICA), and detail its benefits for automotive applications. Its simple, lightweight design makes efficient use of available resources. Its wide bandwidth allows it to receive multiple services and enables it to continue to function despite frequency shifts introduced by automotive paints. The PICA antenna may enable vehicle manufacturers to remove numerous service specific narrowband antennas, reducing vehicle weight, complexity and wiring requirements. Its planar structure reduces aerodynamic drag, and it's compatibility with low cost substrates allows integration with Sheet Moulding Compound (SMC) or other composites used for vehicle body panels, providing a further reduction in vehicle weight and associated environmental benefits.

1. Introduction

Our generation has witnessed the beginnings of a new era in wireless technology. Wireless systems are now central to both the business world during work hours and the social world outside of work. The automotive industry is not isolated from this wireless revolution. Substantial increases in the use of vehicular wireless systems have occurred in the past decade and the trend is expected to continue into the foreseeable future.

Many decades ago, a simple protruding "mast" type antenna was the only antenna required in a vehicle. It was used for the purposes of receiving analog radio, in both the AM and FM frequency bands, commonly located around 1 MHz and 100 MHz respectively.

In recent years, we have seen new types of antennas fitted to vehicles to accommodate new wireless services that have found application in the automotive environment. GPS navigation systems are now found in many vehicles, and require antennas to receive the

J. Wellnitz et al. (eds.), *Sustainable Automotive Technologies 2010*,
DOI 10.1007/978-3-642-10798-6_6, © Springer-Verlag Berlin Heidelberg 2010

signals broadcast from the GPS satellites. In North America, Satellite Digital Audio Radio Service (SDARS) enables passengers to enjoy a wide selection of high quality audio channels which are also broadcast by a group of satellites.

In addition to these services, the radio capability requirements of automobiles will likely increase further in the future. Vehicles may be fitted with equipment that will operate in additional bands to cater for wireless internet via WiMAX or UMTS HSPA technology, enabling the vehicle to receive XML traffic updates, it's occupants to access email and internet while on the go, and also enabling realization of Location Based Services when used in conjunction with appropriate position information. Location Based Services assist the vehicle's occupants by enabling them to search for nearby places of interest, as an example, the nearest ATM or restaurant. Digital Video Broadcasting may also be incorporated into vehicular systems, enabling provision of video entertainment for rear seat passengers. Other wireless services which may be required in future vehicles include collision avoidance radar, and vehicle-to-vehicle or vehicle-to-infrastructure communication. Both of these systems require use of additional antennas operating at new frequencies, with the aim of enhancing road safety.

2. Existing automotive antennas

Vehicles currently in production commonly use a number of discrete antennas in seeking to incorporate the increasing number of wireless services into vehicles. Retractable mast antennas provide for AM/FM radio applications and are commonly mounted on front or rear fenders. "Bee-sting" antennas are a family of newer antennas which also seek to provide for the AM/FM radio requirements. They are commonly shorter than mast antennas, and are often mounted on the roof of hatchback body style vehicles. Antennas incorporated into the vehicle glass have been known for several decades, and may take many forms, but are often created with conductive pastes, and appear to be similar to defogger elements. Only in the last decade has the most recent form of automotive antenna appeared the "Shark-fin" antenna. A shark-fin antenna commonly incorporates antennas for a different range of services than those of the previously mentioned antenna types. A shark-fin antenna usually consists of a cast base, several discrete narrowband antennas for specific services and a covering structure (Hopf *et al.* 2007). Common antenna elements include GPS patch antennas and multi-band cellular phone antennas. These shark-fin antennas adequately provide for many of the higher frequency needs of a modern vehicle, yet there are several disadvantages with existing designs.

Shark-fin antennas are complex and expensive to fabricate, consist of many parts, and are typically quite heavy for their size. They are prone to damage, theft and vandalism. They protrude from the vehicle body, and therefore contribute to aerodynamic drag. They are highly visible, and although they are typically aesthetically pleasing in themselves, they place certain restrictions on the vehicle designers with regards to styling. An alternative approach to vehicular antenna requirements would ideally decrease the component weight, reduce vehicle manufacturing costs, and provide an environmentally sustainable solution whilst elegantly meeting the challenge of antennas for the increasing number of wireless services.

3. The PICA antenna solution

Rather than affix a specific antenna to vehicles for each narrowband service, or collocate numerous narrowband antennas under a shark-fin radome, a color-matched radiating aperture could be created in the vehicle body using a material such as SMC. If this aperture contained a single antenna of wide impedance bandwidth it may be capable of providing for nearly all the necessary wireless capabilities of the vehicle.

Many wideband antenna designs exist, but the Planar Inverted Cone Antenna (PICA) is of particular interest for this application. It combines a very wideband impedance bandwidth in excess of 10:1, with high efficiency, and is able to perform well on thin, low cost FR-4 substrates.

Several PICA variations have been proposed in the literature since its original publication by Suh et al. in 2004 (Suh *et al.* 2004). Of particular interest for this application is the uniplanar CPW fed PICA slot antenna. In its original form (Pell *et al.* 2008), this antenna occupied a substrate of 120 mm by 120 mm, and provided an acceptable impedance match (> 10 dB return loss) from 1.2 GHz to beyond 13 GHz. This is a very wide impedance bandwidth, being greater than 166% or 10:1, and may provide for many of the required services, yet fails to provide for the lower frequency communication bands.

It is well known that scaling an antenna's dimensions can produce similar performance at a shifted frequency band. With this aim in mind, we increased the antennas dimensions by a scaling factor of 5/3 (or about 1.67) in order to scale down the frequency of operation. The end result is an antenna which occupies a substrate of 200 mm by 200 mm, and is shown in Fig. 1. Detailed antenna dimensions are available in (Pell *et al.* 2009).

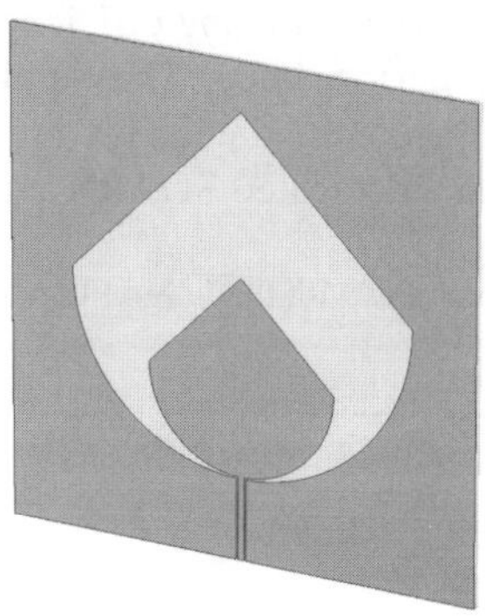

Fig. 1 Perspective view of the Large PICA antenna

3.1 Capabilities of the large PICA antenna

This scaling of the uniplanar CPW fed PICA slot antenna, creates an antenna with impedance match down to 730 MHz, and up to almost 15 GHz (Fig. 2). This wide frequency range covers many of the desired bands in the automotive realm, as shown in Table 1, which range in frequency from 850 MHz to 5.9 GHz, although a separate AM/FM antenna would still be required. This may be achieved by the use of existing on glass designs.

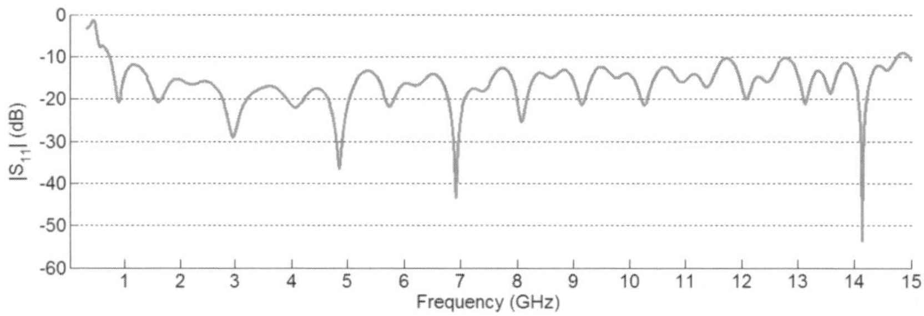

Fig. 1 Simulated impedance response of the Large PICA antenna

Common cellular telephone bands used across the globe are included in the antennas operable frequencies, in the range from 850 MHz to 2.1 GHz. In the author's home country of Australia, Telstra's NextG™ network uses the 850 MHz portion of spectrum to provide the widest coverage of the nation, and is presently the fastest installed HSPA network in the world. Inclusion of this band into the vehicle allows for full realization of Location Based Services, XML based traffic updates from SUNA and internet connectivity almost anywhere across the country.

GPS reception is provided for, although with a loss of 3 dB due to polarization mismatch. Many modern cell-phones implement Assisted GPS and also neglect to provide circular polarization for the incoming GPS signal, yet provide acceptable position information, and we expect that the higher gain and efficiency of the present PICA design will provide equivalent or superior performance. The Satellite Digital Audio Radio Service used in North America is also accommodated, yet is also a circularly polarized service and may be better received with a dedicated SDARS antenna.

Table 1 Summary of Signals Compatible with the Large PICA Antenna

Service Name	Frequency
Cellular Phone	850 – 900 MHz
GPS	1.575 GHz
Cellular Phone	1.7 – 2.1 GHz
SDARS (Satellite Digital Audio Radio Service)	2.3 GHz
Bluetooth, 802.11 b/g/n Wifi	2.4 GHz
WiMAX	3.5 GHz
DSRC, V2V, VII, 802.11p	5.9 GHz

Additional services incorporated include those which occupy the ISM band at 2.4 GHz including both Bluetooth and Wi-Fi, and WiMAX (which may also be used to provide broadband wireless access and is implemented at 3.5 GHz in some areas of Australia). The 5.9 GHz band has been reserved for vehicle-to-vehicle and vehicle-to-infrastructure communication, and is sometimes referred to as Dedicated Short Range Communications or 802.11p. In the near future, it is expected that this band will see the formation of vehicle networks to improve road safety.

Clearly, the 200 mm PICA antenna makes ample provision in it's impedance response for the aforementioned frequency bands in use today, and any applications that may arise in the future in the region up to and beyond 10 GHz. However, impedance response is just one of the performance measures of an antenna. Radiation direction and gain must also be appropriate for each desired service. Five frequency bands of interest have been chosen to illustrate the antenna's radiation performance (Figure 3).

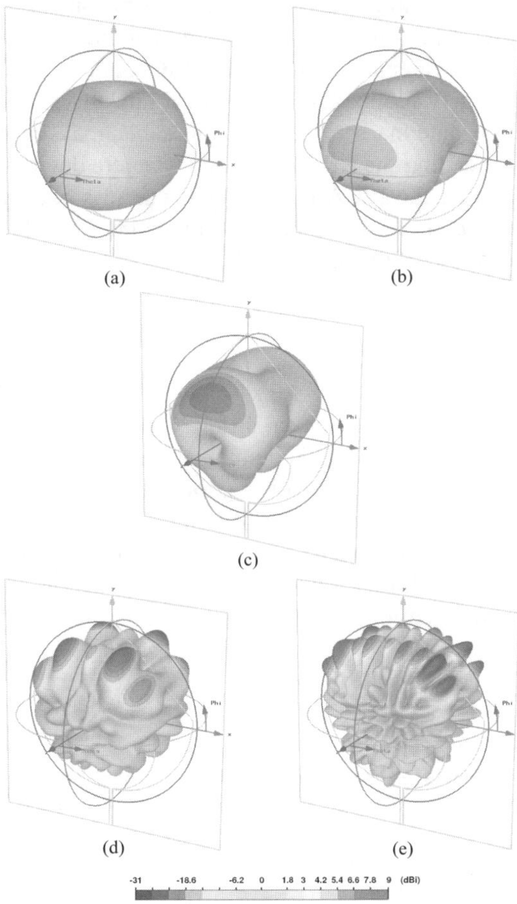

Fig. 3: Simulated 3D radiation patterns of the enlarged uniplanar CPW fed PICA. (a) 850 MHz (b) 1.575 GHz (c) 2.4 GHz (c) 5.9 GHz (d) 9.5 GHz

3.2 Sustainable advantages of the PICA in vehicular systems

The implementation of the PICA antenna in vehicles may provide several benefits over other designs, helping to create sustainable vehicles and an environmentally friendly automotive market. The antenna is lighter and far simpler than shark-fin designs, being fabricated as a single structure. The PICA also makes efficient use of available materials. It uses affordable industry standard FR-4 substrates, meaning the antenna is etched onto a thin layer of copper cladding which covers the substrate. This approach uses less metal than some competing antenna designs. The use of FR-4 also gives manufacturers great freedom, enabling the antenna to be fabricated in any one of a multitude of electronics factories worldwide.

The planar nature of the antenna enables it to be incorporated under polymeric body panels. Benefits of this are two fold. Firstly it eliminates the aerodynamic drag induced by other externally mounted antennas due to their protrusion from the vehicle body. Secondly, it promotes use of non-metallic light-weight composite parts, further reducing the total vehicle weight. The wideband nature of the PICA antenna allows it to receive multiple services, and this may reduce the required cabling in vehicles. Instead of separate coaxial cables for each antenna service, a single cable may be used, with appropriate splitting infrastructure feeding the RF inputs of the discrete devices.

Overall, we expect that the PICA antenna will use less energy and water in vehicle construction, and reduce fuel consumption throughout the vehicle's life span due to weight saving and reduction of the drag coefficient.

4. Conclusion

The need for wireless access in vehicles at an increasing number of frequencies may see future vehicles adopting a single wideband antenna rather than individual service specific narrowband antennas. The PICA antenna is a good candidate for such applications due to its wide bandwidth, and low profile nature. It is fabricated on lightweight materials, and when integrated with composite panels, creates no parasitic aerodynamic drag, improving fuel economy. The PICA's shifting radiation pattern throughout the band facilitates both terrestrial and satellite applications.

References

Hopf J., Reiter L. & Lindenmeier S., 2007. Compact multi-antenna system for cars with electrically invisible phone antennas for SDARS frequencies. In *2nd International ITG Conference on Antennas*. pp. 171-175.

Pell B. et al., 2008. The design and realization of uniplanar CPW fed PICA slot antennas. In *2008 Asia-Pacific Microwave Conference*. pp. 1-4.

Pell B.D. et al., 2009. Custom-Directional Wideband PICA for Multi-Service Vehicular Applications. In *2009 Asia-Pacific Microwave Conference*. Singapore.

Suh S.Y., Stutzman W. & Davis W., 2004. A new ultrawideband printed monopole antenna: the planar inverted cone antenna (PICA). *IEEE Transactions on Antennas and Propagation*, 52(5), 1361-1364.

Part 2
Solution for reduced pollution and congestion

EMISSION VERSUS E-MISSION – WAYS TO A SUSTAINABLE POWERTRAIN

T. Dräxl

Gear Research Centre (FZG), Technische Universität München,
Boltzmannstr. 15, 85748 Garching, Germany
E-mail: draexl@fzg.mw.tum.de

Abstract Powertrain efficiency is one of the main starting points if sustainable mobility should be achieved. Necessary improvements are possible by both optimising conventional drivelines and fuels and introducing entirely new concepts. This paper presents and evaluates measures to enhance fuel economy and to reduce emissions of road vehicles.

1. Meeting the CO2 challenge – efficiency at all fronts

Mobility of people and goods has been a cornerstone of growth and wealth in ancient and modern societies. The spread of the automobile in the 20th century brought individual transport to the masses. Despite the economic crisis the global demand for mobility will continue to grow especially in the emerging markets. Since the worldwide stock of passenger and light duty vehicles is expected to increase from 650 million in 2006 to 1.4 billion until 2030 (OECD/IEA, 2006), large efforts will be necessary to ensure the accustomed mobility for the foreseeable future. Sustainable mobility must be efficient, clean and safe throughout the whole vehicle life cycle, but at the same time it must be affordable. A holistic approach that integrates society, environment and economy is required.

With respect to fuel economy and emissions, the appropriate use of primary energy and the whole process chain through energy conversion and transport and the choice of mobile energy carriers and propulsion systems must be taken into account. Causing about 11 % to 13 % of the global carbon dioxide emissions, road vehicles are not the largest single polluter or energy consumer, but they are a considerable and well-visible one. The European Parliament's Environmental Committee has decided to force the automotive industry to cut CO_2 emissions of new vehicles to 130 g/km until 2015. These European and other emission regulations force carmakers to improve fuel economy of their products significantly or even to offer non-fuel-driven vehicles.

Today the typical means of individual transport is a 4-wheel vehicle that needs about 96 % to 98 % of its total energy demand to move itself and not the driver or passengers. Lightweight construction, aerodynamics and rolling resistance are im-

J. Wellnitz et al. (eds.), *Sustainable Automotive Technologies 2010*,
DOI 10.1007/978-3-642-10798-6_7, © Springer-Verlag Berlin Heidelberg 2010

portant parameters if this transport efficiency should be improved. But since the powertrain is what actually drives a vehicle, it is one of the most important starting points for optimisation.

2. Evolution of established technology

Battery electric vehicles (BEVs) have been arousing interest in the media recently. Manufacturers put into operation prototype fleets and energy companies are installing recharging or even battery exchange stations. One can gain the impression that this is the beginning of a revolution. Mainly due to the limitations of present battery technology, BEVs cannot yet fulfil long-standing customer expectations especially regarding cruising range and anytime availability, however. Internal combustion engines (ICEs) will probably remain the predominant drive configuration for a while and must be further improved. Current trends concerning spark-ignition (SI) engines are downsizing and supercharging, variable valve train and direct injection. Diesel engines still have benefits over them regarding fuel efficiency, but they increasingly suffer from the high exhaust aftertreatment effort necessary to comply with NO_x and particulate matter limitations. Possibly a new generation of ICEs using new combustion processes like homogeneous charge compression ignition (HCCI) will blur the differences between diesel and SI engines.

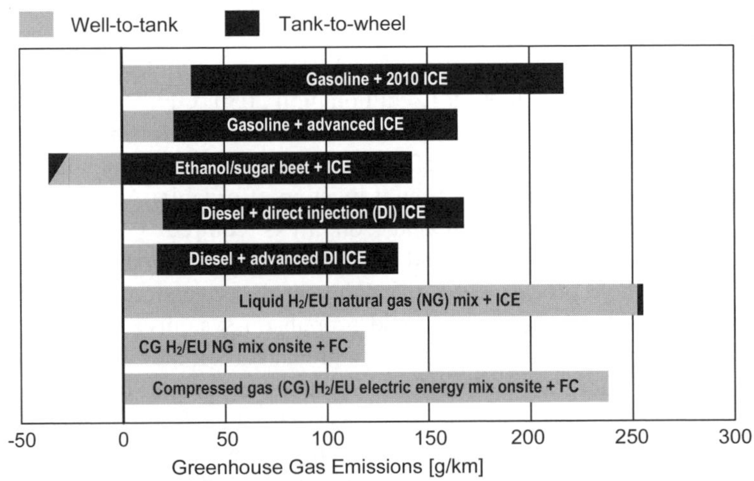

Fig. 1 Well-to-tank and tank-to-wheel (sum: well-to-wheel) greenhouse gas emissions for various fuel and propulsion system combinations (WBCSD, 2004).

An ICE needs a transmission to adjust rotational speed and torque to the needs of the vehicle. The market is dominated by manual (MT), automated manual

(AMT), automatic (AT), double clutch (DCT) and continuously variable (CVT) transmissions, with MTs still achieving the lowest losses. CVTs in contrast allow for efficient operation of the ICE. Measures to reduce losses in transmissions include optimised choice and application of bearings, lowering the oil level, use of low-viscous lubricants, baffle plates to reduce the accelerated oil volumes and replacing wet clutches by dry clutches to eliminate hydrodynamic drag torque. Even gears, although being very mature machine elements, can be further improved with respect to efficiency. "Low loss" gears are characterised by small module/pitch, high pressure angle, high face width and small transverse contact ratio (Höhn *et al.*, 2005) and can reduce losses in the tooth contact by more than 50 %.

New fuels will be required to deal with limited resources although currently mineral oil based fuels are still the cheapest choice. Bio-fuels are gaining in importance, but due to commercial and social considerations their possible production volume is very limited, too. The energy input required to produce gasoline or diesel typically amounts to about than 20 % of the usable chemical energy contained in the finished fuel (Schmidt, 2008). Bio-fuels tend to require a higher energy input, the plants used as feedstock absorb CO_2 during growth. **Fig. 1** gives a brief overview of the emissions caused by fuel production and fuel use.

Hydrogen as a mobile energy carrier used in ICEs or fuel cells could be a viable alternative to batteries if the appropriate infrastructure and storage technology are provided. With present high pressure tanks, an effective energy density in the scale of 1 kWh/kg is possible. This is far below the heating value of the hydrogen itself, but still by one order of magnitude better than present batteries. Together with the possibility of fast refuelling, this makes hydrogen interesting to extend the cruising range of BEVs.

3. The electric (r)evolution?

BEVs feature high in-vehicle (battery-to-wheel) efficiency and offer the possibility of driving quietly and without local (tail-pipe) exhaust emissions. Actual well-to-wheel emissions can only be determined by assessing carbon intensity of electricity generation, however. These emissions can be higher or lower than for conventionally driven vehicles, depending on the energy mix of the region considered and other model assumptions. A McKinsey (2009) study expects general CO_2 benefits for electric vehicles over fuel-driven ones from 2017 on if electricity generation is optimised accordingly. An ambitious, but conceivable scenario in this study assumes sales of 8 % EVs, 24 % hybrid EVs (HEVs), 28 % plug-in HEVs (PHEVs) and 40 % optimised ICE-driven vehicles until 2030. This could yield a 49 % CO_2 reduction compared to a "no measures at all" scenario and 22 % compared to the 2006 emission levels despite the growth of the vehicle stock.

Energy storage plays a key role for electric propulsion. Present NiMH and Li-ion batteries reach a maximum energy density of about 140…170 Wh/kg. The actual energy density of vehicle batteries is still lower (NiMH battery of the Toyota Prius II: 46 Wh/kg) because sufficient service life and dynamic behaviour must be achieved, and trade-offs between energy storage capacity and power capability are required. Significant further improvement will require new material combinations like e.g. metal-air batteries. Further challenges of battery technology include monitoring of state of charge and charge distribution, dynamic charge acceptance throughout the operating temperature range, safety issues for vehicle applications and of course cost. Today, various sources state costs for a passenger car traction battery between 250 € and 1,200 € per kWh capacity. For applications requiring fast charging and discharging rather than high capacity, double layer capacitors (supercapacitors, ultracapacitors) can be an alternative to batteries or can also complement on them in combined storage solutions.

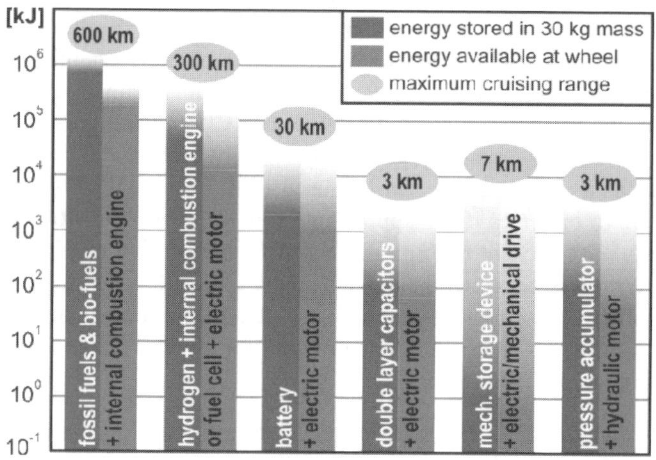

Fig. 2 Stored and usable energy for several propulsion systems (Höhn & Dräxl, 2009).

In Fig. 2 a comparison of the amount of energy contained in 30 kg of various fuels and regenerative storage systems (tanks and casing included) is shown. A rough estimate of maximum attainable cruising range in a mid-size car is given. Even if BEVs could reach competitive cruising ranges, the question of extending the range remains. Since recharging times in the scale of minutes seem unrealistic even on the long term if both electrochemistry and power networks are taken into account, only complex battery exchange systems or range extenders using ICEs or fuel cells come into consideration. A far future vision could be a wireless inductive energy supply on the road. Large research and development efforts will be required to resolve the unanswered questions in this field.

Hybrid propulsion is a much discussed issue, and unlike BEVs, hybrid electric vehicles (HEVs) have already gained a notable market share. About 541,000 HEVs were sold worldwide already in 2005 (Kahn Ribeiro *et al.*, 2008). A reasonable hybrid system combines the benefits of ICEs and electric drives while avoiding their drawbacks. This is possible by measures such as ICE start-stop, electric power-assist in combination with downsizing of the ICE, regenerative braking and ICE load point shift. Temporary electric driving can allow for local emission-free operation. Especially for diesel hybrids, electric modulation of transitions to "phlegmatise" the ICE could be a promising way to meet future emission requirements without compromise on vehicle dynamics.

A key question when designing a hybrid driveline is how much electric power and energy storage capacity to install. A typical large family car requires relatively little power (less than 30 kW) to complete common standardised driving cycles such as the NEDC. On the one hand, this means that the IC engine runs in the inefficient, low partial-load range most of the time. On the other hand a large portion of the theoretical fuel savings potential is accessible with a small electric machine. Hybrids are often classified into micro, mild and full hybrids. Micro hybrids increase fuel economy in stop-and-go traffic through engine start-stop and can provide rudimentary regenerative braking via the alternator. In urban driving cycles they are capable of fuel savings in the single digit percent range despite their comparatively low complexity. Mild hybrids allow for electric power-assist and regenerative braking with a traction motor dimensioned smaller than the IC engine, whereas full hybrids can provide electric driving with a certain range.

A number of hybrid driveline topologies are possible. In a serial hybrid the ICE powers a generator. Electric power is transferred to one or several electric motors that drive the wheels directly or through a gearbox. The ICE's operating point can be freely chosen. However, the electric motor(s) must be dimensioned for the full nominal power of the vehicle and losses tend to be higher than in a purely mechanical driveline. For passenger cars, such a configuration might make sense for range extender concepts with a small ICE. A parallel hybrid is a structure in which both the IC engine and usually one electric motor can drive the vehicle together or independently from each other. Only one electric machine is necessary. Parallel hybrids can be derived comparatively easily from virtually any existing transmission concept and are a viable solution for mild and full hybrids. Some of the currently available series production hybrid cars use a power-split driveline topology with two electric machines and a planetary superposition gear set.

A drawback of hybrid drivelines is that more components than in a conventional or battery electric vehicle must be merged into one car, and as for BEVs, electric energy storage is still a bottleneck. Since it does not make sense to combine two expensive technologies to form one unaffordable system, synergies are to be found by e.g. simplifying ICEs for the specific needs of hybrids. However, costs will probably fall faster for hybrid/electric components than for established technologies such as ICEs.

An online survey conducted by the German Fraunhofer Institute for Systems and Innovation Research (ISI) among 190 powertrain experts shows that mild and micro hybrid concepts are expected to reach a 5 % market share until 2015. The predominant opinion on when full hybrids would take hold varies between 2015 and 2020. PHEVs and BEVs are expected to conquer the market only after 2020. There is little agreement about fuel cell driven vehicles: 19 % of the respondents stated that fuel cells would never play an important role for automotive propulsion, whereas 38 % expect them between 2020 and 2030, and another 15 % later than 2030. Although it is impossible to reliably predict what individual mobility will look like in 50 or 100 years, we must realise that there is no silver bullet powertrain that meets all requirements.

4. Conclusions and perspectives

Are hybrids a pioneering technology for introducing purely electric traction on a grand scale or are they an inadequately complex interim solution that should better be completely skipped on the way to other alternative propulsion systems? Probably the energy demand for transport will have to be met by a greater variety of primary sources. This will also require more different propulsion systems. For urban use electric and plug-in hybrid/range extender vehicles seem promising. For highway use and heavy vehicles, diesels and new ICE concepts, partially with alternative fuels, will remain the way to go for the foreseeable future. Diversity will also be necessary for different markets, since megacities pose different requirements than areas with long driving distances. For complexity and cost reasons, this diversity should not be exaggerated, however.

A high energy and emission savings potential is technically accessible today, but trade-offs are required between what is theoretically possible and what makes sense for marketable vehicles. In any case, fuel and emission saving measures will cause additional costs. The dilemma about this is that the higher the fuel cost, the less money remains to buy cars with a better environmental impact. Conclusions from this are that a high priced, innovative powertrain should offer additional benefit beyond efficiency if possible and that tax benefits or subsidies may be necessary to support renewal of the vehicle stock.

It must be kept in mind that decisions taken today, such as developing new powertrain concepts, will have consequences for decades. It depends on these decisions if we will find less, smaller and more expensive cars leading to limited personal mobility or if there will be an increased demand for more efficient vehicles. This is also an engineering challenge, and renouncing mobility does not help anyone if we can beat this challenge.

References

Höhn B.-R. and Dräxl T. (2009): Hybrid Cars – A Preliminary Solution or Future Perspectives? In Bartz W. J. (Ed.) *Fuels 2009: Mineral Oil Based and Alternative Fuels; 7th International Colloquium.* Esslingen, Germany, January 14 – 15. TAE, Esslingen.

Höhn B.-R., Michaelis K. and Wimmer A. (2005): Low Loss Gears. Paper given at the *AGMA Fall Technical Meeting.* Detroit, 16 – 18 October.

Kahn Ribeiro S., Kobayashi S., Beuthe M., Gasca J., Greene D., Lee D. S., Muromachi Y., Newton P. J., Plotkin S., Sperling D., Wit R. and Zhou P. J. (2007): Transport and its infrastructure. In Metz B., Davidson O. R., Bosch P. R., Dave R., Meyer L. A. (Eds.) *Climate Change 2007: Mitigation. Contribution of Working Group III to the Fourth Assessment Report of the Intergovernmental Panel on Climate Change.* Cambridge University Press, Cambridge, UK.

McKinsey & Company, Inc. (Ed.) (2009): *Roads toward a low-carbon future: Reducing CO_2 emissions from passenger vehicles in the global road transportation system.* McKinsey, New York City.

OECD/International Energy Agency IEA (Ed.) (2006): *World Energy Outlook 2006.* OECD, Paris.

Schmidt G. (2008): Energy Storage and the Path to Sustainable Transportation. In Lenz H. P. (Ed.) *29. Internationales Wiener Motorensymposium.* Vienna, 24 – 25 April. VDI-Verlag, Düsseldorf.

World Business Council for Sustainable Development WBCSD (Ed.) (2004): *Mobility 2030: Meeting the challenges to sustainability. The Sustainable Mobility Project. Full Report 2004.* WBCSD, Conches, Switzerland.

Part 3
Sustainable vehicle design and manufacture

BENCHMARKING AND OPTIMISATION OF AUTOMOTIVE SEAT STRUCTURES

M. Leary, M. Mazur, J. Gruijters, A. Subic

RMIT University, Bundoora, Victoria 3083, Australia
Phone: +61 3 9925 6177, E-mail: martin.leary@rmit.edu.au

Abstract Despite the long history of automotive design and legislation regarding occupant safety, there appears to be little design convergence in the implementation of automotive seating structures. This observation is likely due to compromise between conflicting design requirements including: envelope of motion, structural integrity, modularity, product cost and the cost and complexity of redesign and validation. A range of seating architectures are identified and compared to provide insight into these conflicting design requirements and to aid design optimisation.

1. Introduction

Automotive seating structures are subject to a demanding set of comfort and safety requirements. Comfort requirements are dictated by the customer and include: occupant positioning, seat interface comfort, memory functions, response time and cost. Safety requirements include government legislation for occupant safety in high and low speed impact scenarios, as well as insurance agency expectations of repair costs. In response to these varied and often conflicting requirements automotive manufacturers have responded with a range of automotive seating structures – each of which embodies a compromise between the associated design requirements.

This work presents a benchmarking study of contemporary automotive seating structures. These structures have been categorised according to the associated kinematics, driving mechanism and degrees of freedom. Based on these attributes, the relative performance has been estimated for each identified embodiment, including: achievable range of motion, response time and an estimate of structural integrity. This outcome assists automotive manufacturers to apply a systematic approach to automotive seat design and optimisation based on a robust evaluation of existing embodiments.

J. Wellnitz et al. (eds.), *Sustainable Automotive Technologies 2010*,
DOI 10.1007/978-3-642-10798-6_8, © Springer-Verlag Berlin Heidelberg 2010

2. The design problem

Automotive seating structures provide the interface between the user and vehicle. These structures are required to accommodate a range of users with significant variation in shape and mass. The seat may be studied as a combination of two rigid bodies: the base and back with the following Degrees of Freedom (DoF):

- Translational DoF (fore-aft, lateral and lift)
- Rotational DoF (yaw, roll and tilt)
- Rotation between the seat base and back

Only a selection of the potential DoF are implemented in mainstream commercial vehicles: Fore-aft and lift translation is always implemented; and, tilt is often, but not always, implemented. Additional DoF such as variable seat-base length is present only in few high range vehicles. A range of kinematic solutions are available to implement these identified DoF. Solutions implemented in contemporary vehicles were identified by reference to commercially available benchmarking images (A2mac1.com) and evaluated according to the: achievable range of motion, response time and estimated structural integrity.

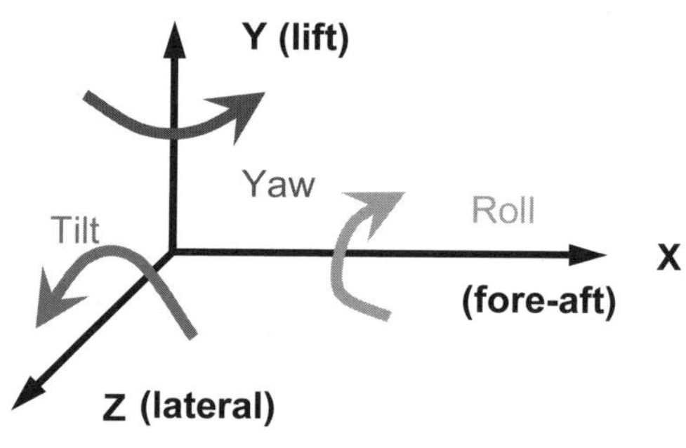

Fig. 1 Definitions of the 3 rotation for a rigid body.

2.1 Fore-aft translation

This motion is used to position the user with respect to the controlling devices (pedals and steering wheel) and in the observed implementations was always independent from of any other DoF. Fore-aft motion is always generated with a sliding rail system, which typically consists of two similar rails, but may be a perpendicular rail system (Figure 2). Fore-aft translation may be implemented manually with an incremental toothed arrangement or powered by a lead screw or rack and pinion (Figure 3).

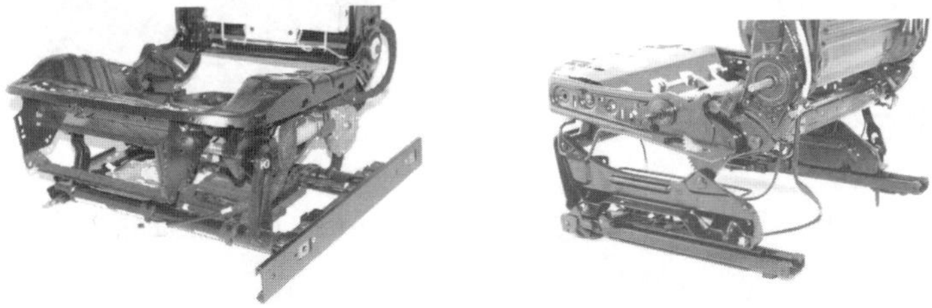

Fig. 2 Fore-aft translation. Left: perpendicular rail. Right: parallel rail.

Fig. 3 Fore-aft mechanism. Upper: Rack and pinion: Lower: Screw drive

2.2 *Vertical translation*

This motion is used to position the user with respect the controlling devices (pedals and steering wheel) and allow visibility. This DoF is typically implemented with a four-bar linkage controlled by crank or sector gears. This implementation results in some unavoidable fore-aft translation, and can incorporate some tilt (Figures 4 and 5).

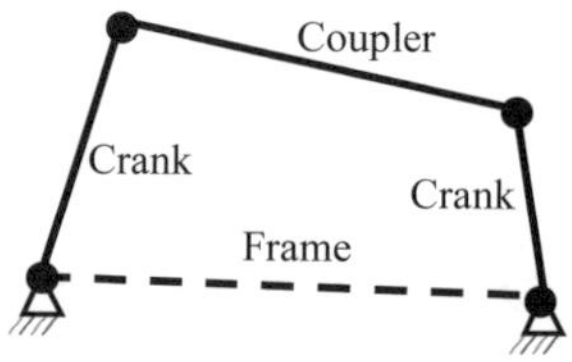

Fig. 4 Four-bar linkage. Left: terminology. Right: Commercial implementation.

2.3 Tilt rotation

Tilt allows the seat base to be aligned with the occupant's thigh to distribute pressure evenly in a natural seating position. Tilt may be actively implemented and/or may be integrated with vertical translation. Alternate implementations are: independent rotation and the front and rear of the seat base; and the addition of a two-bar bar linkage on the four-bar linkage coupler (Figure 5).

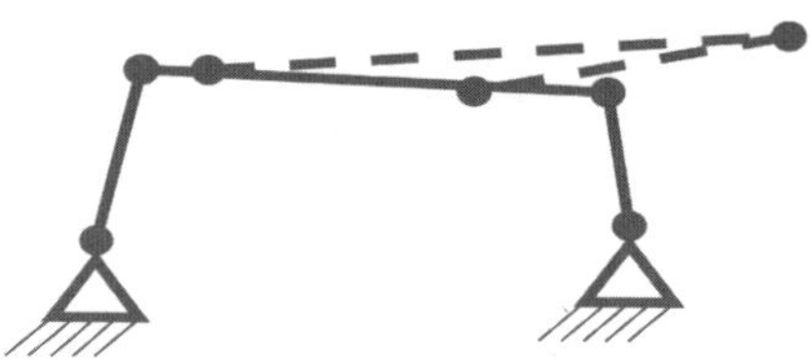

Fig. 5 Two-bar linkage. Left: Schematic. Right: Commercial implementation.

3 Simulation

A series of identified vehicle seat implementation were obtained and the associated geometries were reverse engineered (Figure 6). This data was used to generate a series of virtual models which were assessed to provide insight into: achievable range of motion, response time and estimated structural integrity. For all implementations a quasi-static load of 1127N was applied to represents a 95kg occupant and 20kg supported seat mass. The load is applied at the estimated center of mass: 45 % of the length between the rear and front pivot and 95 mm above rear pivot. Simulations were completed for quasi-static loading, and no conclusions about dynamic behaviour can be drawn.

Seat A

Four-bar link lift actuated by crank at seat rear.
Two-bar link tilt (not shown).

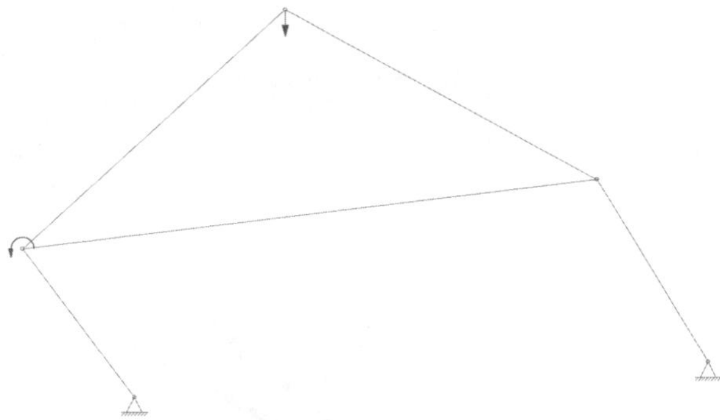

Seat B

Four-bar link lift actuated by crank at seat rear.
Tilt integrated with four-bar link rocker at seat front.

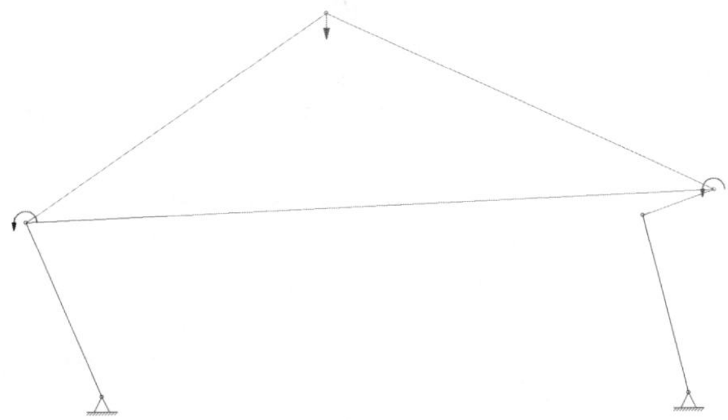

Seat C

Independent actuation of rocker at front and rear of seat base.

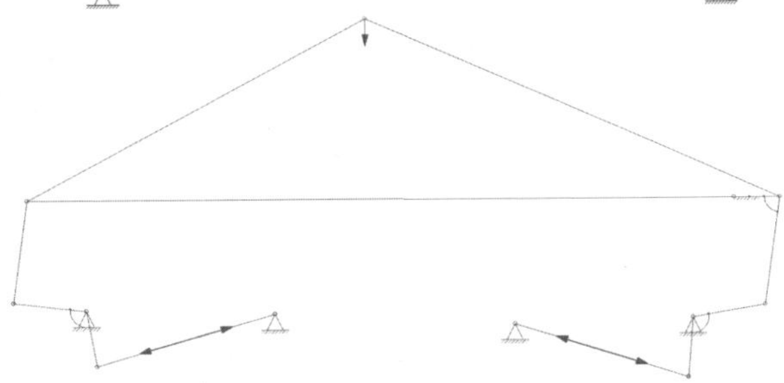

Fig. 6 Seat implementations assessed.

3.1 Lift translation

For all implementations a quasi-static load of 1127N was applied to the estimated center of mass – this represents a 95kg occupant and 20kg supported seat mass. 45 % of the length between the rear and front pivot is the length from the rear pivot of the seat to the x-coordinate of the point mass (Figure 7).

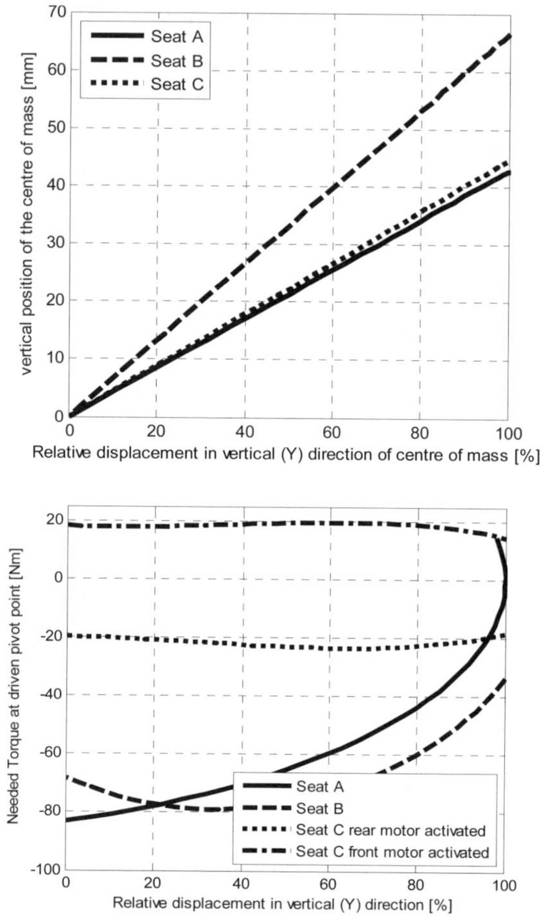

Fig. 7 Upper: lifting height. Lower: required torque.

3.2 *Tilt rotation*

The tilt of the seat is changed by two different actions: Lift of the seat and tilt of the seat. As shown at the left in figure 8 the tilt of seat A and B is changed by lifting the seat, while seat C doesn't change much. For tilting the seat only seat B and C are considered, because for seat A the tilting pan wasn't modeled. Seat B can only tilt in one direction, while seat C can tilt in both directions. The top line of seat C is the torque needed at the front pivot and the bottom line of seat C is the rear pivot of the seat. With a negative relative angle the rear motor is tilting and vice versa.

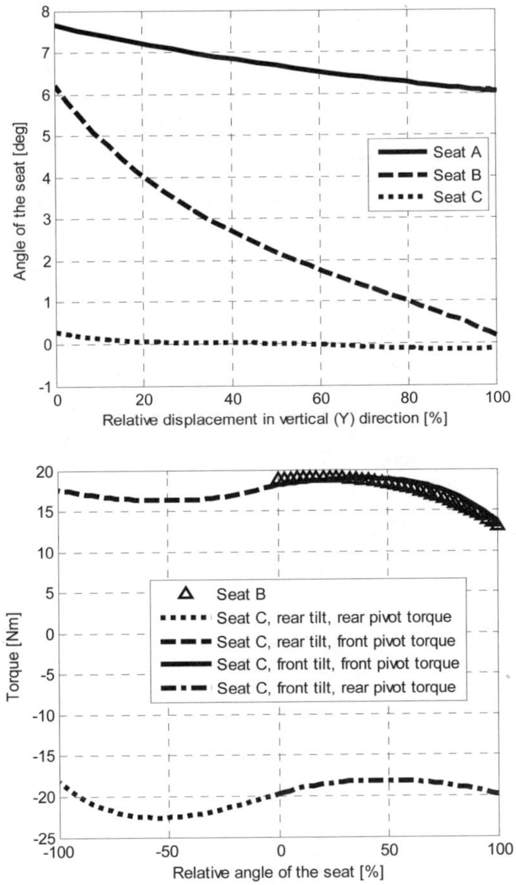

Fig. 8 Left: seat angle. Right: required torque.

3.3 Safety

The most damaging legislated impact scenario for automotive seating is rear impact. To simulate international design requirements a moment of 530Nm was applied about the seat frame pivot in a rearward direction. The resulting tensile link forces (Figure 6) are plotted versus relative vertical height (Figure 9). Positive normal forces are tension forces, negative normal forces are compressive forces.

Seat A
Maximum tension torque is in bar 2 and 3, the maximum compression torque is in bar 1.

Seat B
Maximum tension forces are in bar 2, maximum compression forces are in bar 1.

Seat C
Maximum tension forces are in bar 15 and 5, maximum compression forces are in bar 1.

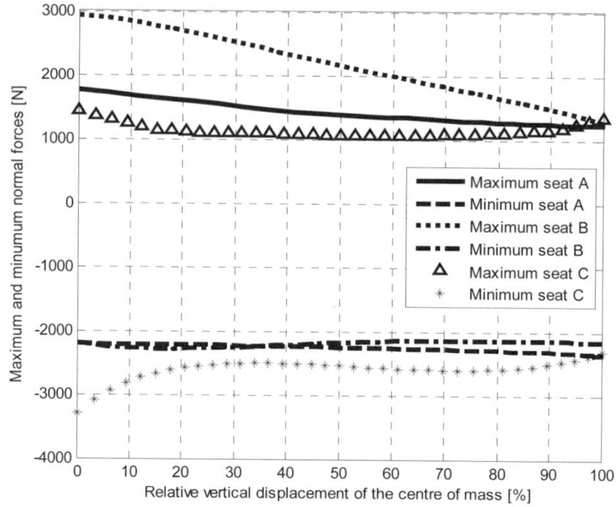

Fig. 9 Tensile link forces.

4 Outcomes

The reverse engineering of seat implementations allows the rapid generation of a range of data on the relative performance, due to space restrictions only a small sample of which has been summarized here. Fundamental outcomes are:

- Despite the variance in seating implementations, the DoF implemented are consistent, i.e. fore-aft, lift, tilt and recline.
- Four-bar linkages are typically (but not always) used to achieve lift. These implementations display an increasing seat inclination with lift to accommodate shorter users who naturally require a combination of higher seating position and additional tilt.
- The peak torque required to achieve lift of Seat C is approximately 50 % lower than for Seat A and B, but 2 actuators are required. The peak torque is relatively constant for all identified implementations (Figure 7).
- Peak link forces during simulated impact provide an estimate of the fundamental structural integrity (Figure 9). Compressive loads on long links are especially important as they are susceptible to buckling. Tensional loads, as exist in the bell-cranks of Seat C, should also be considered.

VISION-AIDED POSITION CONTROL METHOD FOR MANUFACTURING MACHINES

C. A. Montes[1], J. C. Ziegert[2]

[1]International Center for Automotive Research, Clemson University
Clemson, South Carolina, USA, E-mail: carlosm@clemson.edu

[2] Professor & Timken Chair in Design
International Center for Automotive Research, Clemson University
Clemson, South Carolina, USA, E-mail: ziegert@clemson.edu

Abstract A new high-precision multi-axis position feedback approach for manufacturing machines is presented. The desired displacement command of, for example, a cutting tool in a planar 2-DOF environment is emulated by an active element or active target on a liquid-crystal display (LCD). A digital sensor incorporated in the control system of the tool images the active target on the LCD screen, enabling the motion of the active element to be detected and tracked over time with high accuracy. The in-plane physical distance between the actual position of the tool and the target is used as the position error in the control loop. Based on the error signal, the servos that control the motion of the axes are driven accordingly to reduce the position error back to zero. The proposed method is implemented on an X-Y stage and is aimed to be utilized in CNC machines for mass-production parts. Experimental results show that accuracy levels on the order of 5 μm can be reliably achieved.

1. Introduction

In the automotive industry a lot of resources, both economic and human, are being invested in the creation of new technologies that have a less negative impact on the environment, and yet achieve the same or better performance levels of the components used today. A good example is the construction of more efficient internal combustion (IC) engines. Friction between individual parts is one of the main obstacles engineers try to address in these machines in order to more efficiently transform chemical energy into mechanical energy. It is in this area that accurate part manufacturing plays a key role in the development of sustainable technologies, not only in the automotive industry, but also in the aviation and aerospace industries. Most of the time, this translates into higher tolerance requirements enforced by the OEM, which must be met by the suppliers.

Computer numerical control (CNC) machines have been widely used in the past to produce parts in large quantity. In a multi-DOF environment the motion of

J. Wellnitz et al. (eds.), *Sustainable Automotive Technologies 2010*,
DOI 10.1007/978-3-642-10798-6_9, © Springer-Verlag Berlin Heidelberg 2010

a CNC machine is monitored by a control system, which collects information provided by several sensors. Rotary and linear encoders, as well as interferometers are common position sensors found in these machines. The proposed approach uses a digital imaging sensor for multi-axis position control. Instead of computing the position error for individual axes based on axis specific sensors, the new system determines the error based on the in-plane physical distance between a known target imaged through the digital camera and a reference point in the image plane. Implementation is carried out on an X-Y stage, and the target is generated dynamically on an LCD screen. An image processing algorithm for the target detection as well as the control system scheme are explained.

2. Testbed

In most CNC machines, the position of the tool is determined through sensors that do not follow the device directly, but compute the position in an indirect matter. The rotating angle of a leadscrew, or the actual displacement of the sensor along a linear scale are common measurements used by the controller to estimate position. This approach is subject to different external errors, such as axis-alignment errors, part deformation due to temperature variations, and machine wear among others (Chan *et al.*, 2008). Figure 1 shows a typical mechatronic configuration used for axis control. As seen the desired trajectory in a 2-DOF system, $(x,y)_{des}$, is determined outside the control loop and so is the error compensation; hence, the controller cannot compensate for these errors.

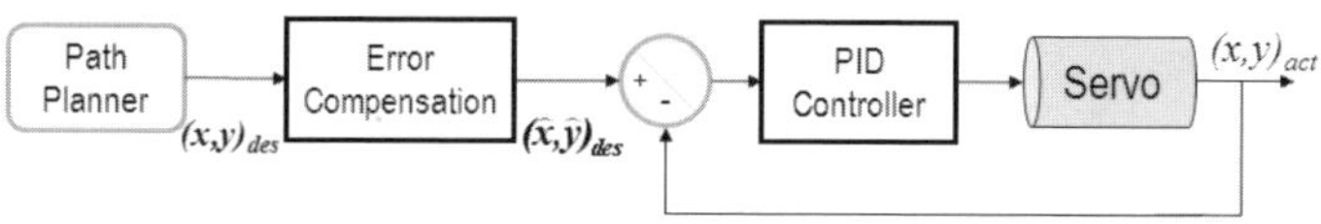

Fig: 1 CNC machine control loop.

3.1 X-Y stage

The current research is implemented on an X-Y stage that allows 2-dimensional motion. An LCD screen is placed upside-down on top of the stage and moves with the stage (Figure 2 (a)). Although displacement is achieved through ordinary leadscrews, connected to servo motors, the position of the stage is determined through a stationary CCD digital camera that is fixed to the support structure of the stage and aimed at the LCD (Figure 2 (b)). If a target image is displayed on the LCD screen, then the position of this target with respect to a reference point can be identified in the image plane of the camera (CCD). Figure 2 (c) shows the control loop configuration resulting by introducing the LCD and imaging sensor into the system. Notice that the position of the stage is no longer measured indirectly through rotary or linear encoders, but it is measured directly with respect to the position of the target.

Fig. 2 (a) LCD mounted on top of X-Y stage, (b) Stationary digital camera, (c) Modified position control loop.

3. Signal processing

In the X-Y stage-setup described before, motion commands are provided through a moving active target displayed on an LCD screen. The position of the target on the LCD is controlled by the user through software. If the principal point PP on the CCD (Figure 3 (b)) is used as the reference, then a position error vector can be defined at any time *t* as *e(t)*, with respect to this reference. For example, if the target shown in Figure 3 (b) was purposely located at point *C* on the LCD, then the corresponding error vector would be *e(t)=c* on the CCD. This vector can be used as the control loop position error, where the controller's goal is to bring the target to the reference position, i.e. PP=(*0,0*).

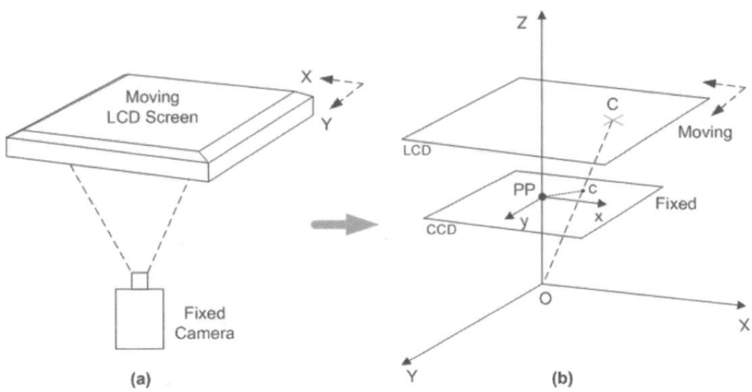

Fig. 3 (a) Imaging sensor looking at LCD, (b) pinhole camera model.

Since high resolution is desired, the construction of the target on the LCD must be studied. Figure 4 (a) shows how one LCD pixel looks when inspected through a microscope with a 90x magnification. As expected, three color bars or stripes corresponding to (right to left) Blue, Green and Red (BGR) are used in the construction of a single pixel.

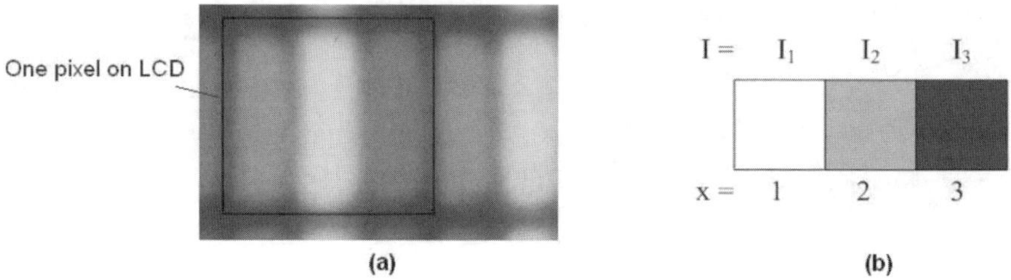

Fig. 4 (a) 90x magnification of one LCD pixel, (b) Single pixel intensity analysis.

The structure of the LCD pixel suggests that the smallest variation on the active target, disregarding color intensity variations, is equal to 1/3 of a pixel on the LCD. Now, if the grayscale intensity, as captured by the imaging sensor, is considered along with the object's location for position control, a feature combining both of these characteristics must be defined. A practical solution arises by computing the intensity weighted center of mass, or centroid (Montes *et al* 2009). If the image $I_{mxn}(x,y)$ is analyzed row-wise where $x \in [0, m-1]$, then at any given row, e.g. $y=j$, the center of mass is defined as

$$X_{C,j} = \frac{\sum_{x=0}^{m-1} xI(x,j)}{\sum_{x=0}^{m-1} I(x,j)} \quad (1)$$

Due to the fact that the wavelength (color) information is lost during the photons-to-electrons transformation performed by the CCD, the camera is unable to actually capture color. Nevertheless, the individual bars (BGR) forming a single pixel on the LCD can still be used to generate different intensity patterns, which can be analyzed using (1) in the CCD. Consider three stripes associated to a single pixel on the LCD screen, with positions x=1,2,3 and assume that the color information is not important and only their grayscale intensities, I_1, I_2 and I_3 are studied (Figure 4 (b)). Furthermore, assume that the grayscale on the LCD screen is quantized in N intervals, with a step size of Q. Direct evaluation of (1) gives the intensity centroid

$$X_{C0} = \frac{I_1 + 2I_2 + 3I_3}{I_1 + I_2 + I_3} \quad (2)$$

The minimum intensity change in the third stripe (read from left to right) results in a new centroid, namely

$$X_{C1} = \frac{I_1 + 2I_2 + 3(I_3 + Q)}{I_1 + I_2 + I_3 + Q} \tag{3}$$

From (2) and (3), the centroid variation for the minimum intensity change can be expressed as $\Delta X_C = X_{C1} - XC0$. ΔX_C represents the position resolution on the LCD screen. If 8 bits are used to span the entire grayscale, then the smallest centroid variation, i.e. the best achievable resolution, can be obtained by setting I_1=I_2= I_3=255 and letting I_3 vary from 255 to 255-Q. That is

$$Res_{LCD} = |\Delta X_C| = \left| \frac{3*255 + 3*(255 - Q)}{2*255 + (255 - Q)} - 2 \right| \tag{4}$$

Given that the same intensity quantization can be sensed by the camera, the position resolution on the CCD is calculated using the pinhole camera model ([Chan *et al.*], 2008). Note that a point C on the LCD screen with coordinates (X,Y,Z) in the coordinate system shown in Figure 3 (b), maps to a point $\boldsymbol{c}$ on the CCD with coordinates $(x, y) = (fX/Z, fY/Z)$, where f is the focal length of the lens. Applying the same model used for point C, the resolution in the CCD is computed,

$$Res_{CCD} = \frac{f}{Z} Res_{LCD} \tag{5}$$

4. Target Detection

The position of a known target on the LCD is determined and tracked by the control system over time using the algorithm proposed in (Montes *et al.* 2009). The target is represented by the intersection of two diagonal lines as shown in Figure 5 (a). The horizontal location of each point on the diagonals is obtained by calculating the intensity weighted center of mass around the two intensity peaks on each row of the digital image (Figure 5 (b)). The diagonals are calculated through linear regression and the intersection of the functions (target) is easily computed analytically.

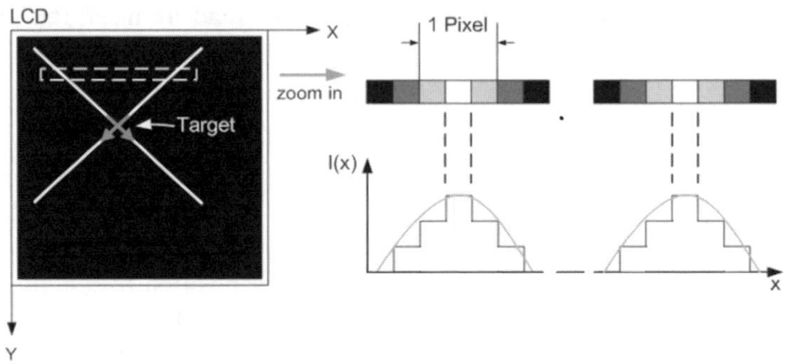

Fig. 5 (a) Dynamic target on LCD, (b) row-wise individual location of the points in the diagonals trough centroid.

5. Experimental Results

The ability of the vision sensor to detect displacement is tested. Figure 6 (a) shows the horizontal displacement recorded using a rotary encoder versus the same displacement as capture by the camera, when the stage is operated under normal closed loop control. For each data point associated to the imaging sensor, a 20 images sample is collected and the final value is obtained by averaging the sample. The minimum displacement change measured through the rotary encoder (also confirmed using an interferometer) is 1.25μm. Due to magnification, this displacement maps to a horizontal displacement of 0.1025 μm in the imaging sensor. For horizontal displacement the image processing algorithm presented a standard deviation of 0.078 μm, which is slightly bigger than half of the minimum displacement detected (0.1025/2=0.0513 μm). Hence, it can be concluded that further data processing is required to detect horizontal displacements smaller or equal to 1.25μm using the proposed setup. However, experimental data confirmed that displacements along the X-axis of 2.5 μm are reliably detected through the proposed approach. A similar analysis for vertical displacement showed a best achievable resolution of 5 μm for this type of motion. The difference between the horizontal and vertical resolutions is due to the way a single LCD pixel is constructed.

A second test is conducted to determine the smallest displacement command that the user can successfully enter through the LCD-camera configuration. If the intensity of the target was not considered, the smallest displacement command that the user could enter would be represented by shifting all the points in the target by one pixel in the same direction on the LCD. Figure 6 (b) shows the change in the target location as a function of the intensity change on the red and green stripes of the pixels on the target. In this case a minimum displacement change of 0.8894 μm is detected through the camera, with a standard deviation of 0.0403 μm. Since the standard deviation is smaller than half of the minimum displacement changed detected by the camera, it can be stated that displacement commands of approximately 0.88 μm can be generated through intensity variations on the target pixels.

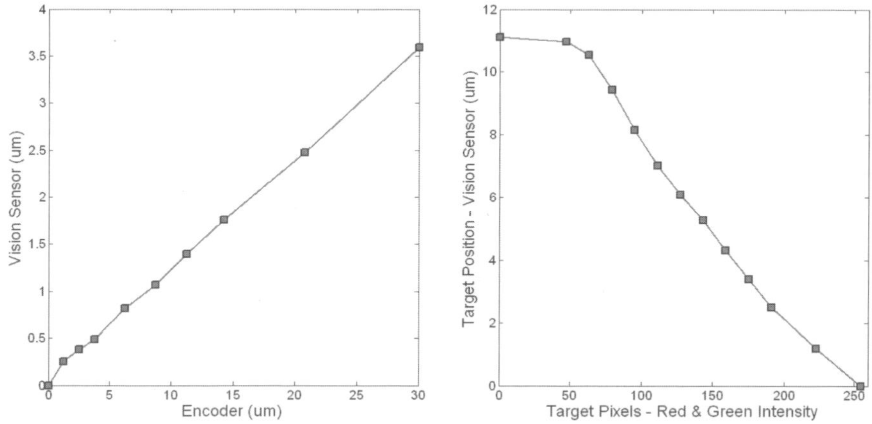

Fig. 6 (a) 30 μm horizontal displacement measured using a rotary encoder and a vision sensor. (b) Target position change due to a simultaneous intensity change on the red and green stripes on the target pixels on the LCD, when the grayscale is quantized into 16 intervals.

6. Conclusion

A multi-axis position feedback method was presented. Experimental results showed that resolutions in the order of 5 μm can be reliably achieved through active target detection, when intensity variations are not considered. Additionally, pixel intensity variations on the target pixels allow displacement commands as small as 0.88 μm. On the other hand, several challenges must still be addressed if the proposed procedure is to be used in real manufacturing machines. Higher accuracy requirements and lower time delays in the control loop are clear issues that this configuration still faces.

References

Lyons L.S., Liu J., Sutton M.A. (1996). “High-temperature Deformation Measurement Using Digital Image Correlation”, Exp. Mech. 36, 64–70.

Montes C.A., Ziegert J.C. and Mears L. (2009). “Method to Measure Planar Displacement using Centroid Calculation”. 37 NAMRC. Greenville, SC, USA.

T.S. Newman and Jain, A.K (1995). “A system for 3D CAD-based inspection using range images” *The Journal of the Pattern Recognition Society,* V. 28, No10, pp1555 1574.

Wong C., Montes C.A., Mears L. and Ziegert J.C. (2008). "A New Position Feedback Method for Manufacturing Equipment." ASME, Evanston, IL, USA.

AUTOMATED FIBRE PLACEMENT WITH IN SITU ULTRA-VIOLET (UV) LIGHT CURING: CONCEPT TESTING

A. Di Pietro, P. Compston

School of Engineering, College of Engineering and Computer Science
The Australian National University, Canberra ACT 0200, Australia
E-mail: adriano.dipietro@anu.edu.au

Abstract This paper presents preliminary experimental results from an automated fibre placement (AFP) process for the manufacture of a lightweight fibre reinforced composite. The AFP machine is comprised of a placement head on an industrial robot. It lays dry fibre and liquid resin, and the resin is cured in-situ through exposure to ultraviolet (UV) light. The results include machine limit tests; degree of cure testing in in-situ cured bulk resin samples and assessment of composite quality using optical microscopy.

1. Introduction

Fibre-reinforced polymer composites are structural materials that offer high specific strength and stiffness compared to metals and alloys. The use of composite for large automotive parts such as body structures would significantly reduce vehicle weight. This would improve fuel efficiency, and hence reduce harmful exhaust emissions. Reduced weight would also increase the viability of new green power sources that currently have a lower energy density. Glass-fibre composites are attractive for use in automobiles due to their low cost, availability in large volumes and potential for reduced environmental burden (Puri *et al.*, 2009).

Composite material manufacturing is characterised by inherently expensive and complex processing. Even by employing the most efficient methods currently available in industry, composites cannot be manufactured as cheaply as competing materials. This means a product built from composites will cost more than one built from, for example, plastics or metals. Consequently, composites are often only manufactured for products whose unique material requirements or attributes justify the expense; eg. aerospace structures.

There are three key problems that make composite manufacturing expensive and time consuming; the high cost of constituents which are often in a pre-impregnated (prepreg) form, the time consuming, labour intensive manufacturing process, and thermal treatment in an oven or autoclave to achieve full resin cure.

J. Wellnitz et al. (eds.), *Sustainable Automotive Technologies 2010*,
DOI 10.1007/978-3-642-10798-6_10, © Springer-Verlag Berlin Heidelberg 2010

Automated fibre placement (AFP) could address these three problems. It is a fully automated composite manufacturing process that uses industrial robots to make composite manufacturing more cost-effective, safe and efficient (Bannister, 2001). AFP machines apply composite constituents onto a mould to produce composite parts. During the placement process, multiple fibre tows are pulled from a spool or supply reel and directed to the application head at the end of an industrial robot arm, where they are collimated into a single band, combined with liquid resin and pressed onto the mould of the part. The applicator head consolidates the fibre and resin as they placed in discrete and defined paths. Parts are built up ply by ply.

AFP produces higher quality parts than conventional manual composite manufacturing processes with repeatability that can be ascertained to less than 0.3mm. The use of robots improves accuracy and repeatability of fibre alignment, while allowing the designer to tailor the composites internal structure and performance properties through precise placement of fibres in high-stress areas. For these reasons, the aerospace industry has adopting AFP in order to increase manufacturing throughput time on new composite parts.

The AFP machine for the current study is a prototype from a previous concept study (Di Pietro *et al.*, 2008) and is comprised of a placement head on an industrial robot arm, as shown in Figure 1. It uses inexpensive dry fibre tows and liquid resin, instead of prepregs. Laying dry fibres not only addresses the cost issue associated with prepregs but allows users to manufacture dry fibre performs for liquid moulding processes, such as resin transfer moulding (RTM). Through control of the fibre/resin ratio a user can create a structure with very low resin content (the resin only acts as a binder). Laying the ‘dry’ fibre onto a mould means that the formed shape is preformed as per the designer’s specification. As well as dry fibre performs, by delivering the resin in liquid form and not as a prepreg, the designer has freedom to select alternate fibre-resin combinations and is not restricted to currently available prepregs. AFP with in situ ultra-violet (UV) light curing has potential to consolidate many steps of the composite layup process and could eliminate the need for autoclave or thermal post curing. This would reduce overall manufacturing times. Also, by eliminating the need for thermal treatment in an oven or autoclave, size restrictions on a single composite part could be removed (Bannister, 2001).

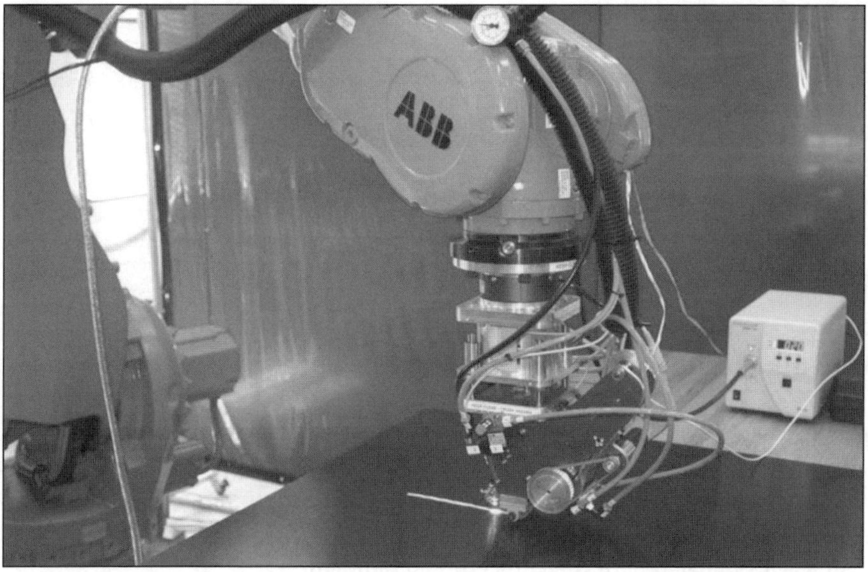

Fig. 1 Automated Fibre Placement with in-situ UV curing of glass-fibre/vinylester composite.

2. AFP with in-situ UV curing: machine specifications tests

The first tests performed on the AFP machine shown in Figure 1 were operational limit tests. The results are presented in Table 1. The ability to operate the machine over large ranges was explicit to the design, allowing for the greatest range during process optimisation.

Each of the specifications presented were determined experimentally to verify the design. The consolidation force range was measured using a 6 degree of freedom load cell integrated into the AFP machine design. Resin flow rates were measured over the operating pressure range of the pneumatic pressure system that provided the positive pressure to push the resin up to the spray nozzle. This range was 0 to 6 bar. Resin viscosity characteristics were determined by varying not only the temperature of the resin but also the styrene content, however numeric viscosity values were not obtained. The UV curing system was verified using a radiometer measured at the same distance the light was positioned on the placement head from the mould surface the resin was sprayed on. The data in table 2 represents comparable operating range with the speeds, resin flow rates and light intensity theoretically allowing consolidated fibre and resin to be laid and cured in a continuous process.

Table 1 AFP machine specifications

Specification	Metric
Consolidation force	10 – 1500N
Resin flow rates	0.3 – 30ml/sec
UV radiation intensity	0.01 – 11.5 W/cm^2
Layup speeds	0.5 – 500 mm/sec
Composite	Glass-fibre/vinylester (Derakane 411-350)

3. Bulk Resin and In-situ UV Curing Tests

The first layup tests included only liquid resin, delivered by an automated spray system linked to a controller and in-situ curing. This step was aimed at testing the resin delivery and cure subsystems. A high intensity UV light was passed over vinylester resin as it was sprayed onto the mould in layers, as shown in Fig. **2**. The resin was formulated with Irgacure 819 photoinitiator.

A significant technical challenge in laying resin and curing it in-situ is the delivery of the resin in a controlled manner. By using a pneumatically actuated spray nozzle with the resin pressurised behind the piston, spraying could be switched on and off on demand. Numerous parameters affected the laid resin characteristics. Varying the nozzle spray size and the pressure behind the resin the flow rate could be accurately controlled. The speed of the layup was also important in controlling the lay down of the resin. By matching the robot speed with the projection speed of the resin as it exited the spray nozzle, the resin remained on the surface in a precise bead similar to the width of the fibre tow. Further, by varying the styrene content of the resin the viscosity could be controlled. Varying the temperature of the resin also affected the viscosity of the resin and how it was delivered. Conflicting considerations in this setup were that the resin viscosity needed to be low enough to be sprayed and delivered to the sprayer, while it was desirable to also form a bead on the mould. To control this, the resin was heated within the pressure pot and the mould was chilled so that the resin formed a bead when it contacted the mould.

The bulk resin samples were tested to determine their degree of cure and to verify the in-situ curing approach. The total UV radiation required to effect cure is characterized by the dosage, measured in J/cm^2. This is a product of the two main processing parameters, the UV intensity (or irradiance) measured in W/cm^2 and the exposure time. A dosage in the range of 7.5-12 J/cm^2 can be sufficient to cure relatively thin (~0.8mm) composites (Crivello *et al.*, 1997). Much of the research on UV curing composites to date has been conducted at relatively low dosages in the order of mW/cm^2. Exposure times in the range 10-30 minutes would therefore have been required (Compston *et al.*, 2008). However, the AFP machine in the current study work used a high intensity (25 W/cm^2) UV light source and reduced the required exposure time to only several seconds. Previous work (Di Pietro & Compston 2009) has shown that a dosage of 375 J/cm^2 is required before the hardness of vinylester resin is the same for both the incident and non-incident surfaces of an 8 mm thick sample. Hence a similar dosage was applied here.

Fig. 2 Placement of the glass fibre tow showing UV light exposure after consolidation.

Initially, the bulk resin layers were exposed to 11.5 W/cm^2 and laid at a speed of 10 mm/sec with a exposure area of 20 mm in diameter. This meant that each unit area of resin was exposed to 23 J per pass. The resin flow rate for each ply was approximately 0.5 ml/sec and the sample was comprised of 6 layers and produced a sample 8mm thick. To ensure all layers in the sample received at least a dosage of 375 J/cm^2 the robot arm speed was reduced to 1mm/sec for the lay down of the final layer.

The resin hardness was characterised according to ASTM D 785-89 and used as a qualitative indication of the degree of cure of the resin. The average hardness on Rockwell Scale M was 53, with a standard deviation of +3.3. This value is lower than that reported in previous work for UV cured and thermally post-cured vinylester (Compston *et al.*, 2008). However, this hardness value is double the value recorded for a vinylester resin cured for 24 hours at room temperature. This result indicates that UV curing in this AFP process has potential to produce composites with a reasonably high degree of cure compared to the common room temperature cure method. Further work will investigate the processing parameters required to achieve a hardness value of 80-90, which is the typical range recorded for the post-cured vinylester sample.

4. Composite placement tests

Fibre and resin samples were manufactured following the successful layup of bulk resin samples cured in-situ. The fibre and resin samples were laid into a strip mould to contain the flow of resin as the constituents were consolidated. The sample contained 10 plies, with a resin flow rate of 0.5 ml/sec and a layup speed of 20 mm/sec. The sample was exposed to 11.5 W/cm^2 of UV radiation energy with an exposure area of 20 mm as with the bulk resin sample. On the final pass the speed for this sample was also slowed however this time to only 5mm/sec to ensure full cure. It appeared that the inclusion of glass fibres assisted in the penetration and absorption of radiation energy as the sample achieved comparable hardness to the bulk resin sample with less exposure.

Fig. **3** shows the cross section of the composite. The resin-rich interlaminar layer can be seen, which is typical of laminated composites, and there are also a couple of voids. There is good wet-out of the fibre by the resin, and a typical fibre distribution pattern with good degree of consolidation and uniformity can be observed. The fibre volume fraction is estimated to be in the range 30-40%. The ideal placement parameter settings, such as consolidation pressure and placement speed, required to eliminate voids and achieve a target fibre volume fraction will be determined in future. Nevertheless, the results shown here indicate that a composite of acceptable quality can be produced with the current AFP process based on dry-fibre and liquid resin delivery, and rapid in-situ UV curing.

5. Conclusions

This paper presented the first experimental results for a robotic AFP process that uses dry fibre, liquid resin and in-situ UV curing. The process shows potential for reducing raw material cost, reducing the number of processing steps and eliminating the need for autoclave post curing. Uniform fibre wet-out was observed and a reasonable degree of cure in the resin was achieved with separate delivery of liquid resin and dry fibre. A partially closed strip mould was used in this work to limit resin overflow, however further work will be conducted to obtain acceptable composite quality using open moulds, and at increased placement speeds. This may require some modification of resin formulation and viscosity.

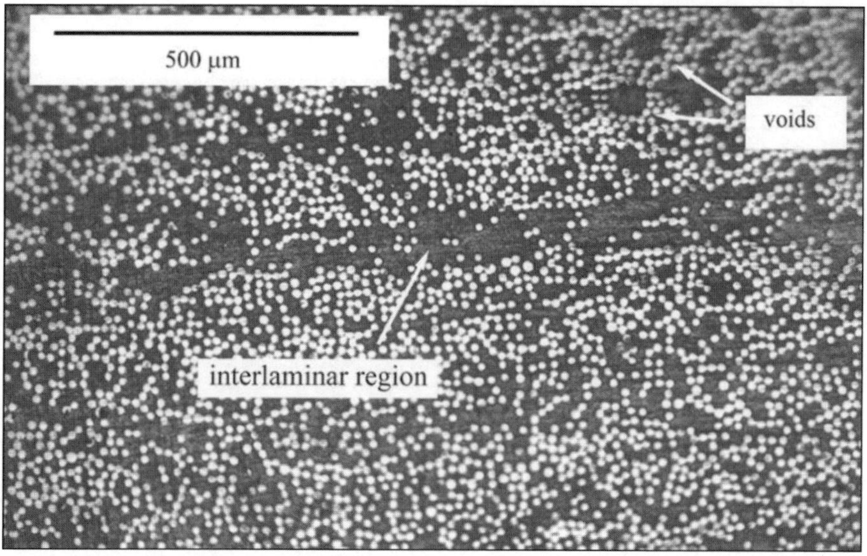

Fig. 3 Optical micrograph of composite cross-section.

References

Puri P., Compston P. and Pantano V. (2009) Life cycle analysis (LCA) of Australian automotive door skins. *International Journal of Life Cycle Analysis*, 14, p. 420-428.

Bannister M. (2001) Challenges for composites into the next millennium – a reinforcement perspective. *Composites Part A*, 32, p. 901-910.

Di Pietro, A., Compston, P., Gresham, R., Hovland, G. (2008) Rapid robotic fibre placement with in-situ ultra-violet curing. In: Subic A., Leary M. and Wellnitz J. (Eds.) *Meeting the Challenges to Sustainable Mobility*. RMIT University Publishing, Melbourne, p. 73-78.

Crivello, J., Narayan, R., Sternstein, S., (1997) Fabrication and Mechanical Characterization of Glass Fiber Reinforced UV-Cured Composites from Epoxidized Vegetable Oils. *Journal of Applied Polymer Science*, 64, 2073.

Compston P., Schiemer J., Cvetanovska A. (2008) Mechanical properties and styrene emission levels of a UV-cured glass-fibre/vinylester composite. *Composite Structures,* 86, 22-26.

Di Pietro A., Compston P. (2009) Resin hardness and interlaminar shear strength of a glass-fibre/vinylester composite cured with high intensity ultraviolet (UV) light. *Journal of Materials Science*, 44, p. 4188-4190.

MULTIFUNCTIONAL VEHICLE COMPONENTS: KEY IN SUSTAINABILITY AND LONG TERM VIABILITY OF AUTO INDUSTRY

E. Sulic[1], B. Pell[2], S. John[1], R. Gupta[3], W. Rowe[2], K. Ghorbani[2], K. Zhang[4]

[1] School of Aerospace, Mechanical & Manufacturing Engineering, RMIT University, GPO Box 2476V, Melbourne, Victoria 3001, Australia

[2] School of Electrical & Computer Engineering, City Campus, GPO Box 2476V, Melbourne, VIC 3001, Australia.

[3]School of Civil & Chemical Engineering, RMIT University City Campus, GPO Box 2476V, Melbourne, VIC 3001, Australia.

[4] School of Mathematical & Geospatial Sciences, City Campus, GPO Box 2476V, Melbourne, VIC 3001, Australia.

Abstract Use of sheet moulding compound in the area of manufacturing of vehicle body components for both structural and non-structural applications has been increasing over last 20 years (Sara *et al.* 2003), driven by Original Equipment Manufacturers and their need for lighter more fuel efficient vehicles. In addition, the use of composite materials such as SMC allows vehicle designers to use more complex component shapes that are not feasible with traditional materials such as stamped steel and aluminium. In parallel with this trend the number of entertainment and communication systems in the vehicles has also been expanding. Current services such as GPS, digital radio, and cellular phone, and future services such as collision avoidance radar, vehicle-to-vehicle and vehicle-to-roadside communications require use of antennas to send/receive signals that are necessary for these services to function. This trend has resulted in significant progress in antenna design and their placement in vehicles, having manifested itself in antenna placement between laminated sheets of glass and 'shark-fin' design used by European vehicle manufactures. Development of one single wideband antenna that is capable of receiving all of the services listed above would represent significant advantage for any OEM and their customers. Namely these benefits would manifest themselves in form of lower overall design, reduced vehicle weight and manufacturing vehicle cost for the OEMs and hence, lower fuel consumption for the end users.

J. Wellnitz et al. (eds.), *Sustainable Automotive Technologies 2010*,
DOI 10.1007/978-3-642-10798-6_11,

1. Introduction

Thermoset composites reinforced with high performance fibers such as glass, carbon, aramide and number of natural fibers are used in many advanced applications where weight, modulus, and strength are valued. These include multiple aircraft structures, automotive components, marine hulls, wind energy blades, and many other applications. The total US market for composite materials topped 4 billion pounds in 2004, and has been increasing at a 6.5% annual rate for over 40 years (MacNeil 2007). Composites, though generally more expensive than traditional materials such as steel and aluminium, are used because they deliver benefits that cannot otherwise be obtained. Such benefits are the results of features such as light weight, modulus, high strength, low dielectric constant, and good chemical resistance. Often, the end-uses application demands a combination of properties that eliminate the use of other materials. For example, carbon fiber is used to reinforce epoxy thermosetting resins when the primary features needed are light weight and high strength, such as in the skin of an aircraft or a race vehicle. While other composites reinforced with glass fibers such as unsaturated polyester commonly referred to as SMC (Sheet Moulding Compound) brings advantages such as relatively light weight, medium cost, easy processing and good weight to strength ratio. These types of composites are commonly used in manufacture of structural and class-A automotive components. They offer advantage of light weight, complex shapes and surface finishes, good durability and subsequently reduced fuel consumption when used in vehicle construction.

In order to further enhance superiority of the composites it is feasible to integrate additional functionality in such components, one of them being a communication device such as an antenna. Other communication devices such as RFID tags for vehicle identification and tracking are also possible to embed using the same approach as embedding of an antenna. However this paper will focus on antenna as a communication device that is embedded into composite component. Creating a multifunctional vehicle component such as roof, bonnet or a trunk offers significant advantages to the Original Equipment Manufacturers (OEMs) as it significantly reduces whole life vehicle costs ensuring long term viability and sustainability of the automotive sector. In addition these types of new technologies will offer significant advantages to automotive suppliers in competitive global supply market. This paper will discuss the development of multifunctional composite components in addressing the manufacturing viability and hence, long term sustainability of the automotive industry particularly in relation to its vehicle weight, fuel reduction and environmental credentials.

2. Component Development

Number composite panels used in vehicle manufacture is increasing at a steady rate. These components range from doors and bonnets to roofs and trunks. Concurrently antenna types used for automotive applications are also changing from "shark fin" antenna and laminated glass antenna to multi service panel antenna shown in figures 1and 2.

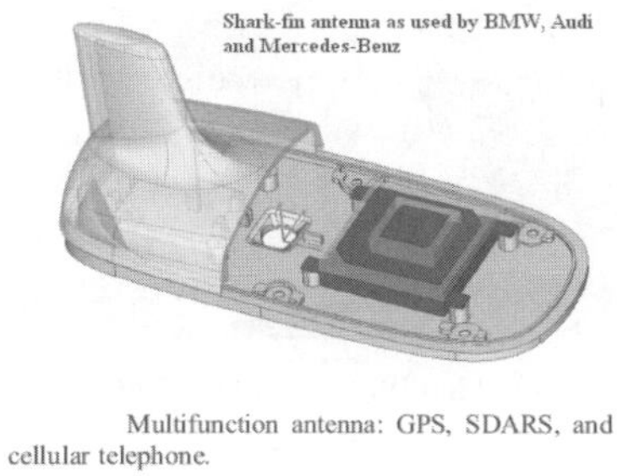

Fig. 1 Shark fin antenna.

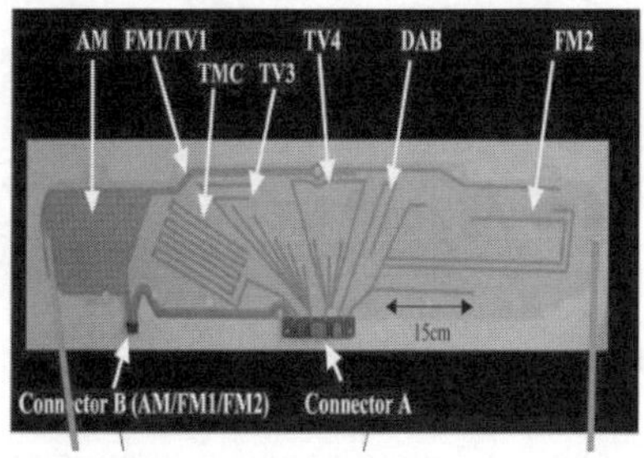

Fig. 2 Multi antenna panel.

However all of the antennas currently used are unique / individual vehicle components that are not integrated into any other vehicle part.

For this project team has developed one ultra wideband planar antenna known as PICA (Planar Inverted Conical Antenna) that is capable of receiving frequencies ranging from about 700 MHz to about 8 GHz. This wide frequency range allows for reception of large number of signals required for following services: Cellular phone, GPS, Satellite digital audio radio services, WI-Fi, Bluetooth, Wimax and future services as vehicle-to-vehicle and vehicle-to-roadside communication at 5.9GHz frequency. This one antenna is therefore capable of replacing number of other individual antennas currently used for each one of the services listed above (Gschwendtner and Wiesbeck 2003). Estimated weight saving in form of wires and individual antennas is from 1.1 Kg to 2.3 Kg depending on antenna locations in the vehicle, size of the vehicle and type of wiring used to connect antennas to the devices that are using the relevant frequencies.

Furthermore once the PICA antenna is embedded into composite panel such as roof of the vehicle additional benefits can be realized. These include assembly time and complexity reductions, assembly labor reduction and additional weight saving due to the use of composite rather than traditional material such as steel.

For purpose of this project a wideband PICA antenna was embedded into Endgate of a "ute" or "pick-up truck" shown in Figures 3 and 4. The resultant weight saving of this composite multifunction system over traditional steel part and individual antennas was 4.75 kg.

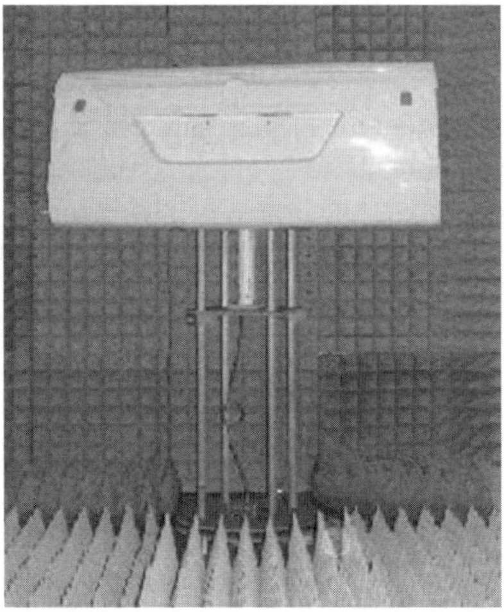

Fig. 3 SMC Ute Endgate.

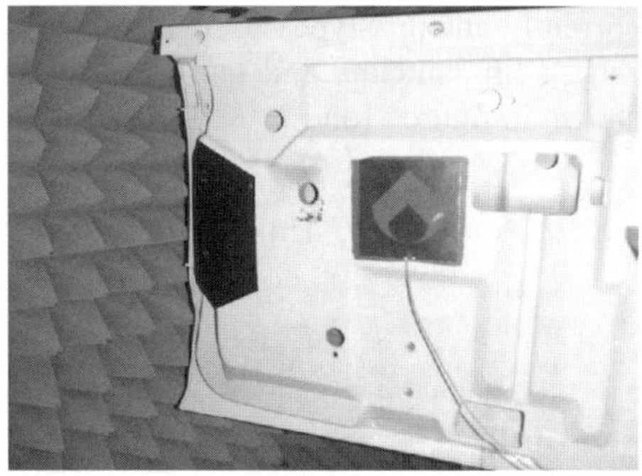

Fig. 4 Embedded PICA into Ute Endgate.

Table 1 below shows a summary of benefits comparing composite multifunctional composite components versus the traditional steel counterparts and individual antennas for some vehicle panels.

Table 1 Weight reduction for some vehicle class-A surfaces when manufactured using multifunctional composites versus traditional steel and individual antennas.

	Total Weight Class-A surface only		
	Traditional material and individual antennas	Multifunctional composite	Reduction in kg
Roof	15.90	10.24	5.66
Bonnet	11.10	6.96	4.14
Trunk	13.02	8.28	4.75
Door	7.90	4.78	3.13

Table 1 input data:

- Specific gravity of steel 8000 kg / m^3.
- Specific gravity of SMC 1750 kg / m^3.
- Material thickness – Steel 0.8 mm.
- Material thickness – SMC 2.5 mm.
- Antenna weight – Individual antennas 0.5 kg.
- Antenna weight – PICA 0.1 kg.
- Wiring weight – Individual antennas 1.5 kg.
- Wiring weight – PICA 0.6 kg.
- Size of individual components for large sedan comparable to BMW 5 series, E class Mercedes or Audi A6.

3. Discussion

In order to be able to embed PICA antenna into composite panel number of challenges had to be resolved. They included:

- Understanding SMC's dielectric constant and its effects on antenna performance.
- Understanding effects of common automotive metallic and nonmetallic paints on antenna performance and signal quality.
- Understanding of compression molding process parameters such as temperature, pressure and material flow and their effects on antenna performance.
- Developing new way of protecting antenna feed lines so that wire can be connected to the antenna after it had been embedded into SMC.

All of the above challenges have been successfully addressed resulting in development of the composite Endgate with embedded PICA antenna. The results for each of the above have been presented at various international conferences and journal papers.

4. Conclusion

The number of vehicle systems requiring use of antennas to send / receive free space signals is on the increase. Next wave of such devises will be headed by vehicle-to-vehicle and vehicle-to-roadside communications at globally reserved frequency of 5.9 GHz. These devices will be used to allow inter-vehicular communication primarily to improve safety and allow drivers to avoid road congestion, effectively converting vehicle into mobile "communication nodes". PICA antenna represents good solution for the above described services as it is ultra wideband, easy to manufacture, has low weight, low manufacturing cost and is planar allowing it to be embedded into class-A composite panel. The advantages of PICA type antennas when compared to narrowband type antennas are very obvious.

The need to reduce vehicle manufacturing and running costs and subsequently life-cycle costs will drive OEMs to develop and use new and innovative approaches to material selection, component design and system / subsystem integration. One of these innovative solutions will include use of multifunctional composite components (Barone and Caulk 1986). Some vehicle manufacturers have already started to bridge the gap between traditional component design and new innovative designs. This is reflected in use of "shark fin" and "multi antenna panel" approaches by European OEM's. Use of multifunctional composite components has a potential to reduce overall vehicle weight by about 18 kg to 20 kg when implemented on vehicle components listed in Table 1. This weight reduction would have significant effect on vehicle's fuel consumption and therefore life-cycle cost and CO_2 emissions.

Acknowledgments This research was supported under Australian Research Council's Linkage Project funding scheme (project LP0561868) and by Composite Materials Engineers (CME) industry partner.

References

Barone, R., M, and Caulk, D.A., "The Effect of Deformation and Thermoset Cure on Heat Conduction in a Chopped-Fiber Reinforced Polyester during Compression Molding", *International Journal of Heat and Mass Transfer*, Volume 22, 1979, pp. 1021-1032.

Barone, R., M, and Caulk, D.A., "Kinematics of Flow in Sheet Molding Compound", *Polymer Composites*, Volume 6, 1985, pp. 105-109.

Barone, R., M, and Caulk, D.A., "A Model for the Flow of a Chopped Reinforced Polymer Compound in Compression Molding", *Journal of Applied Mechanics,* Volume 53, 1986, p. 361

Bird, B, R., Armstrong, C, R., Hassager, O., Dynamics of polymeric liquids, Wiley, 1987, p. 258-263

Castro, J. M., and Griffith, R. M., "Sheet Molding Compound Compression-. Molding Flow", *Polymer Engineering and Science*, Volume 29, Issue 10, 1989.

Castro, J. M., and Tomlinson, G., "Predicting molding forces during sheet molding compound (SMC) compression molding. I: Model development", *Polymer Engineering and Science,* Volume 30, Issue 24, 1990, pp. 1569-1573.

Gschwendtner, E., Wiesbeck, W., "Ultra-broadband car antennas for communications and navigation applications," *IEEE Transaction on Antennas and Propagation*, Volume 51, Issue 8, August 2003, pp. 2020-2027.

Hoare, E.G., Hill, R., "System requirements for automotive radar antennas," IEE Colloquium on Antennas for Automotives, 2000, pp 1/1- 111.

Lindenmeier, H., Hopf, J., Reiter, L., Active AM-FM windshield antenna with equivalent performance to the whip now as standard equipment in car production, Antennas and Propagation Society International Symposium, 1985 p. 621- 624.

MacNeil, R. (2007). *Composites Manufacturing.*

Osswald, T. A., "Numerical Methods for Compression Mold Filling Simulation" Mechanical Engineering, 1987.

Sarah, B., Lisa, M.A., Jose, M. C., Predicting molding forces during sheet molding compound (SMC) compression molding. II:effect of SMC composition, Polymer Composites, 24; 6; 2003 p.731-747

Tucker, C. L., and Liang, E. W., " A Finite Element Method for Flow in Compression Molding of Thin and Thick Parts', *Polymer Composites,* Volume 16, 1995, pp.70-80.

Walbeoff, A., Langley, R.J., Multiband PCB antenna, IEE Proceedings - Microwaves, Antennas and Propagation, 2005 p. 471- 475.

KNOWLEDGE MANAGEMENT IN CROSS-COMPANY RESEARCH AND DEVELOPMENT PROJECTS

S. Lorenz

Private Institute of Technology and Artistic Design (ITD E.V.)
Marie-Curie-Straße 6, 85055 Ingolstadt, Germany
E-mail: sebastian.lorenz@itd-in.de

Abstract Knowledge management methods are suitable to contribute to the success of cross-company research and development projects. Due to globalization, competition and cost pressure research environments are increasing in complexity. To meet development demands of the 21st century companies often connect to project based research networks. Joined forces of team members enable interdisciplinary and complex research approaches within such project organizations. However, project teams need support to manage the knowledge resources. In this article a framework is presented, which can be used to set- up and run knowledge management within such project environments. The framework is based on governing the dimensions strategy, structure and culture. Knowledge management methods and their application have to be adjusted during the project advancement following the focus of the project phases. Emphasis on strategy building is on setting up and synchronizing the knowledge goals. The dimension of designing the knowledge management structure is focused on defining relationship, roles and interfaces within the project organization regarding the knowledge aspects. Implementing the framework for the dimension culture includes enabling accessibility, sharing, exchanging and communication. By applying the described methods, knowledge management can be efficiently installed and can help to lead R&D projects to success.

1. Introduction

Companies who are involved in *research and development* activities are faced with increasing *challenges* to govern their processes. Due to increasing pressure from globalization, competition and cost demands new ways to increase efficiency of those activities are searched for. By joining forces in *cross-company R&D projects*, companies gain advantages by utilizing synergies regarding hardware, human resource and *knowledge*. Cross-company collaboration in R&D topics can have various *advantage*s in terms of knowledge: Access to a broader pool of ex-

J. Wellnitz et al. (eds.), *Sustainable Automotive Technologies 2010*,
DOI 10.1007/978-3-642-10798-6_12, © Springer-Verlag Berlin Heidelberg 2010

perts and expert knowledge, shared usage of expensive equipment, co-use of intellectual property and patents or utilizing networks to increase scientific background and product insights. Such cross-company activities require the set-up of a temporary organization regulating the mode, time and extension of the collaboration. It is widely popular to install temporary *project organizations* to meet the structural needs of the collaborations. The project organization has to build a *suitable framework* for the scientists in order to target activities best. The next level to cross-company R&D excellence will be installing a systematic knowledge management. In the following article an approach to set-up knowledge management in cross-company R&D projects will be described.

2. Knowledge management

The *value of knowledge* as a vital business resource is reflected in a growing number of publications in this area. It is suggested, that we are in the middle of moving into a *knowledge oriented society*. When looking into the meaning of the term "knowledge", the most common explanations rely on defining the relationship between the human action and their underlying information. In order to *decide and act*, humans resort to previous *learning experience*. However, not all "information" can be classified as "knowledge". A continuously advancing quality can be observed from mere data, information which can be built out of it by adding context and the advanced level of knowledge.

Subsequently all actions which are aimed to *manage* the resource knowledge within a company, cross-company network or cross-company project can be summarized as "*knowledge management*". In detail, this can be actions of organizational, human resource, cultural or technical background. Knowledge management *strategies* aim to systematically check, develop, connect, distribute, store or transfer knowledge. A variety of established concepts for company knowledge management exist within theory and praxis. Relevant concepts include for example the "5 phases model" (Nonaka and Takeuchi 1995), the introduction of knowledge management components (Probst et. al. 2003) or approaches applying systemic theory (Wilke 2001). However, concepts for introduction of knowledge management are *not specifically* targeted towards use in *project organizations*. In contrast to knowledge management in companies, cross-company project *blast* both – the *borders of the single organization* as well as the *limit of a static approach*, as a project permanently changes its character following its project phase timeline.

3. Requirements for knowledge management in cross-company R&D projects

In order to design the knowledge management in cross-company R&D projects the *specific requirements* of the concept have to be developed. The general goal of the concept should be to implement knowledge management into R&D projects by *integrating* it as far as possible *into existing project organization* structures. Especially it has to be taken into account that project teams *are cooperating* as virtual cross-company teams and within a highly dynamic environment of research and development. Additionally, it has to be taken into account, that projects are not static, but develop in phases during the *advancement* of time. The requirements can be summarized as followed:

- *Implementation* of knowledge management into R&D projects
- *Integration* into existing project organization structures
- Consideration of *project phases*
- Consideration of *cross-company* issues
- Consideration of *R&D* character

4. Governing knowledge management in projects

Knowledge management requires an interdisciplinary, integrated approach (Haasis and Kriwald 2001). The factors to install a suitable *structure* and establish a compatible *culture* is very important for knowledge management (Tupperinger 2003). Knowledge transfer can be supported by *designing* organizational structures which allow free flow of knowledge and in general avoid hierarchical barriers. Culture is an important point to allow creativity of individuals while at the same time valuing individual knowledge. People have to be supported and encouraged to *share ideas*. Moreover, *strategy* can be identified as a central point to shape a knowledge management environment. Especially for cross-company projects - which have similarities to network oriented organizations regarding the cooperation across company boarders - the definition of the *strategic goals*, the matching of the strategic ideas for the project goals and their implementation is of importance for the *success* of the project (Sydow and van Well 2003). As the list of requirements suggests, an additional dimension has to be the project phase orientation of the knowledge management approach. A project can be divided into *project phases* (see DIN 69901): definition, planning, execution, finalization and the dealing with post project issues. During each of the project phases, specific parameters are of importance. Accordingly, the applied knowledge management methods have to be *adjusted for each phase* to meet the current demands. It can be summarized, that by shaping structure, culture and strategy in dependence on project phases, a framework for integration knowledge management into cross-company R&D projects can be formed.

5. Framework knowledge management in cross-company R&D projects

5.1 Design of strategy

The decision for a *strategy* defines, in which areas, how and with which resources a project will be dealing. With this determination of strategy, a *general decision* about which knowledge will be necessary to fulfill the strategy is taken. As such, *knowledge* becomes a *strategic resource* of a project (Wildemann 2003). This requires the definition of strategic knowledge goals which have to be *connected with the project goals*. The strategic knowledge goals therefore not only describe the status of the knowledge resources at the beginning of the project, but also *outline the goal* for the end of the activities and their planned developments towards it. The knowledge strategy describes how the strategic development goals should be reached (Hinterhuber and Renzel 2002). Important points for planning the knowledge strategy are therefore:

- Definition of the key knowledge for the project
- Definition of a knowledge development plan
- Definition of strategic knowledge goals and connection with the project goals

5.2 Design of structure

Organizational structures need to *support* the project members to perform their knowledge intensive activities. However, defining a knowledge management friendly structure for projects is not easily answered, as each project is embedded in a different environment. On the one hand, research results require freedom *for individual creativity*, on the other hand, too much individualism can result in unconnected knowledge resources which make knowledge transfer between project members difficult (Probst et. al. 2003). As knowledge management requires resources, time and attention, roles and responsibilities have to be assigned. For projects, the *strategic responsibility* for the knowledge management can be assigned to the *project management*. This is where also the responsibility for the project goals and the controlling of the project progress is located. Whereas the *project team members* are the *experts* in dealing with, developing, exchanging or documenting the knowledge.

As a second important part for designing the structural dimension is to set up a *suitable organizational project structure* which fosters information flow. For a holistic knowledge approach, the structure has to map the *relationship* between project management, team members, project partners and external environment. The structure within the project boarders defines the relationship between the knowledge base, consisting of the united knowledge of the team members, the research processes and knowledge management processes. It has to be taken care of, that *appropriate interfaces* enable a free flow of information and knowledge between the entities. This can be done by installing communication channels, reporting and controlling elements or technical instruments for *knowledge exchange.*

5.3 Design of culture

Culture can be a major factor in influencing knowledge management in projects. Many aspects of how culture can boost or hinder knowledge exchange are found in literature. For projects it is especially important, to *enable access of knowledge* across the team members. This comprises, that individuals are *willing to share* their expertise and do not see it as their property which they have to defend. It has to be beneficial and positive to share ideas and concepts with co-workers (Stein 2003). Other important points to be considered are: willingness to *learn*, ability to *handle conflicts* and to work in *a solution oriented* way (Wildemann 2003).

Especially for cross-company projects a *high level of trust* between the project partners has to be established. Project team members need guidance and good example, that it is favored to expose company internal data to project members of a different company, who might be a potential competitor. For R&D projects usually the good example of one of the bigger project partners is required. Trust has to be build, so that team members have *faith in the expertise* of their counterparts. Moreover, it has to be culturally accepted to use, develop or alter ideas of others. It has to be avoided, that such actions are seen as idea burglary or inappropriate criticism.

Efficient knowledge transfer needs a mutual language. Most obvious, team members with different mother tongues need to agree on a common project language. If going more into detail, it is also required to define the use and meaning of specific technical terms. Especially for R&D projects, where not so many previous examples of publications can be found and the appliance of the language terms might be unfamiliar for team members, *glossaries* might be a helpful tool.

5.4 Dependency on project phases

The *timeline* of projects are dominated by the imprint of the *project phases*. As each phase has its own characteristics due to major events, it is obvious to describe the *knowledge management characteristics* accordingly. Figure 1 shows an overview about the project phases and their generic milestones.

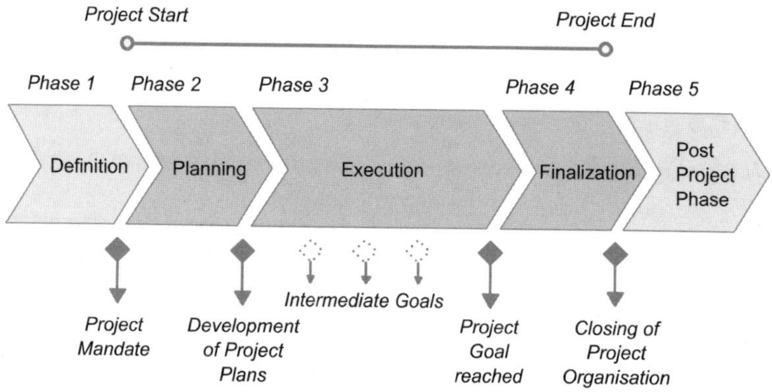

Fig. 1 Project phases and main generic milestones.

As the emphasis of the dimensions "*strategy*" and "*structure*" is to design an appropriate environment for knowledge management, these aspects are of main interest during the *early stage* of the project. Strategic goals have to set during the definition phase of the project. Later on during the *planning phase*, a knowledge development plan and subsequently the organizational structure can be developed. The definition of the roles and responsibilities is best done during the planning phase, as in the project *execution phase* everything as to be set up and operational. The *cultural aspects* have to be considered early in the planning phase regarding a common language, but it is also important to start building a *trustful relationship* between the team members from the *beginning* of the execution phase. Cultural aspects have to be *continuously* monitored and sharing of information has to be encouraged. At the end of the project during the *finalization phase*, the knowledge development gets less important as the main focus is to prepare final reports and develop strategies for *transfer* and *future use* of the research results. Focus during the *after project phase* should be on *justifying* the fulfillment of the strategic goals and from the cultural aspects team members should be *coached* during the transition phase to new projects or back into the company organization. The establishment of networking possibilities even after the active phase of the project can also be beneficial.

6. Conclusion

Research and Development projects are operating in even more *complex* environments these days. For classical R&D projects project members have most likely a highly scientific, R&D orientated background - which not necessarily includes the scopes of managing project, process or knowledge issues. In recent developments *knowledge* has become a very *important resource* in the battle for scientific excellence and steps into the focus of further improving cross-company R&D activities. Knowledge Management can make a *vital contribution* to govern cross-company research and development projects. By designing the *dimensions* of "*strategy*", "*structure*" and "*culture*" in dependence of the *project phases* a knowledge management system can be described for R&D projects. The described framework clearly has *great potential* to systematically shift cross-company cooperation to a next level and improve project effectiveness.

References

Haasis HD, Kriwald T (2001) Wissensmanagement in Produktion und Umweltschutz. Berlin

Hinterhuber HH, Renzel B (2002) Die strategische Dimension des Wissensmanagements. In Bornemann M, Sammer M (Ed) Anwendungsorientiertes Wissensmanagement. Wiesbaden, p. 23.

Nonaka I., Tageuchi H. (1995) The Knowledge-Creating Company. New York

Probst G, Raub S, Romhardt K (2003) Wissen managen – Wie Unternehmen ihre wertvollste Ressource optimal nutzen. Wiesbaden.

Stein W (2003) Best Practice im Wissensmanagement – Ergebnisse einer internationalen Untersuchung und Erfahrungen aus dem Berateralltag. In Hungerberg H, Meffert J (Ed) Handbuch Strategisches Management. Wiesbaden, S. 877.

Sydow J, Van Well B (2003) Wissensintensiv durch Netzwerkorganisation – strukturationstheoretische Analyse eines wissensintensiven Netzwerkes. In Sydow J (Ed) Management von Netzwerkorganisationen. Wiesbaden, p. 108 ff.

Tuppinger J (2003) Wissensorientierter Organisationswandel – Ein Ansatz zur Veränderung von Struktur und Kultur. Wiesbaden, p. 3 ff.

Wildemann H (2003) Wissensmanagement – Ein neuer Erfolgsfaktor für Unternehmen. München, S 6 ff, S. 18ff.

Wilke H (2001) Systemisches Wissensmanagement. Stuttgart

Part 4
New vehicle concepts

STRUCTURAL OPTIMIZATION PROCESSES FOR FINDING ROBUST AUTOMOTIVE LIGHTWEIGHT CONCEPTS

A. Schumacher, K. Hakim

Hamburg University of Applied Science, Department Automotive Engineering and Aircraft Design, Berliner Tor 9, 20099 Hamburg, Germany,
E-mail: schumacher@fzt.haw-hamburg.de

Abstract By designing integrated process chains CAD and CAE come closer together. Throughout the "batch-mode" process chains, CAD-Systems and CAE-Systems for different kinds of simulations (crash, acoustic, …) can be linked. Based on predefined CAD-Parameters different designs can therefore be evaluated efficiently. An additional mathematical optimization procedure can furthermore improve designs without manual intervention. It is useful to integrate commercially available software tools. Therefore, the integration of different CAD/CAE-Tools will be presented. Moreover, using the example of a cast material with inhomogeneities due to the processing, the potential of a fully integrated manufacturing simulation (e.g. cast simulation) will be shown.

1. Introduction

Automotive engineering needs multidisciplinary processes. If we deal with mathematical optimization methods, these processes have to be integrated in a fully automatic scheme. Therefore, the coupling of all involved design and simulation tools is necessary. The direct coupling of highly efficient design tools for complex geometries to numerical analysis tools and simulation software, for safeguarding, is a key element to increase the efficiency of the product development. It is important to know that by coupling the automatic generation of the model is meant. Since the interactions are taking place in "batch mode", a user intervention is not necessary. If such a coupling is implemented, the CAD model generation and the model analysis can be integrated into one optimization process (NOESIS 2008; Schumacher 2005). During the process, the design will be improved by a mathematical optimization algorithm. With respect to the predefined goals and requirements, the algorithm adjusts the user defined CAD parameters. The principle of integrating CAD/CAE and optimization tools will be given and illustrated by some examples.

J. Wellnitz et al. (eds.), *Sustainable Automotive Technologies 2010*,
DOI 10.1007/978-3-642-10798-6_13,

2. Shape optimization based on parameterized CAD models

2.1 Principal scheme

The input files necessary for the simulations are usually created in the batch-mode of a finite element preprocessor. A general script generates for all geometric variations input files without user intervention. Afterwards the simulation results are evaluated automatically. The process can be subdivided as follows:

1. Geometry preparation: The existing geometry of the component must be prepared for the simulation. It is recommended to perform this task in the same CAD environment, in which the model was first designed. Eventually, necessary parts have to be added or redundant parts can be removed. Also, it needs to be tested if common edges of coupled parts are sharing the same line.
2. Creating the finite element mesh using an automatic mesh generator (batch mesher).
3. Material definition (stress-strain correlation, failure criteria, …).
4. Definition of finite element properties, e.g. the thickness of shell elements.
5. Definition of loads and boundary conditions (The batch running definition based on the geometry description is necessary).
6. Assembling of load cases.
7. Solving the problem with the solver.
8. Evaluation of the results in the postprocessor.

Usually, the first 6 steps are realized in the finite element preprocessor, which generates an ASCII based input file for the solver. Flexible CAD models can serve as origin for the generation of simulations models for different disciplines like static, acoustic, crash etc. The CAD parameters describe the topology (position and arrangement), shape, dimension, and the materials of each component. By defining the objective function and constraint function, which are evaluated as parameters from the results of the simulation run, the structural optimization can be started. The main advantage in the integration of the CAD software into the process of structural optimization, is the vast flexibility during shape changes. Additionally, design know-how can be implemented (e.g. via Powercopies, KBE-Tools, …). Furthermore, each iteration produces a usable CAD model. As an example Figure 1 illustrates a simulation sequence with MSC/NASTRAN® and LS-DYNA® managed by Optimus®. The CAD software (Button "CAD_batch") is also implemented. Based on the design parameters x_1 to x_4, Optimus® generates with conditional equations the parameters p_1 to p_5, which are forwarded (via parameter file) to the CAD software. In "CAD_batch" the CAD model is updated and then processed in the finite element preprocessor (Button "FE_pre"). Respective scripts are then generating the input files for the simulation. The finite element postprocessor "FE_post" evaluates the results and exports them into the file "result". Optimus® can then access the results y_1 to y_6. In this sequence the correlation $y_j = y_j(x_i)$ is determined, which is needed for the optimization run.

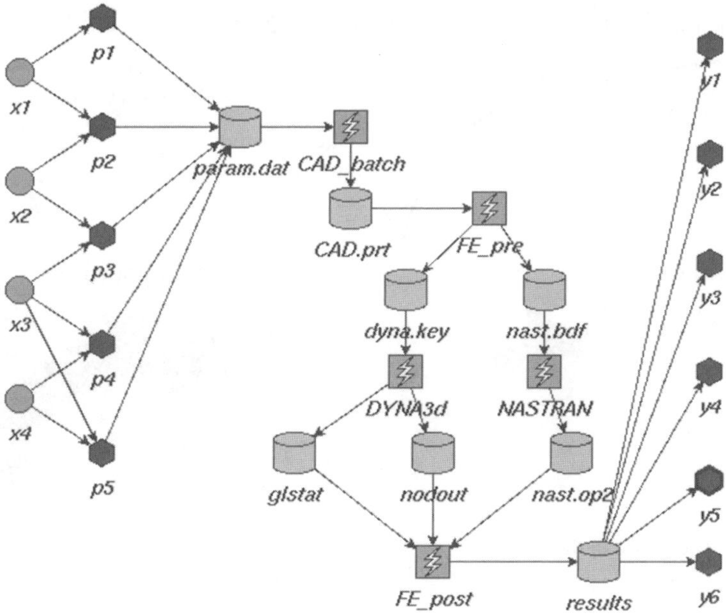

Fig. 1 Example for a simulation sequence in OPTIMUS®.

3. Process chain with line-model based design

At the beginning of a design process when generating the principal structure with a high potential for conceptual changes, line-model based design (Figure 2) is advantageous compared to 'feature'-based design, because line-based models can be generated strictly hierarchical:

1. generating a line-model of the structure,
2. definition of profiles by means of own libraries,
3. providing possibilities for automatic modeling of connecting elements,
4. implementing sheets.

Varying coordinates of control points of the line model provides automatically the changes of all dependent design components. The corresponding FE-model will be also generated automatically. Therefore, changes of the shape result directly in changes of the FE-model (Figure 3), and the implementation of an optimization loop is possible. The software SFE CONCEPT®, which has been programmed to fulfill the special requirements of CAE, provides libraries for different profiles and connection elements.

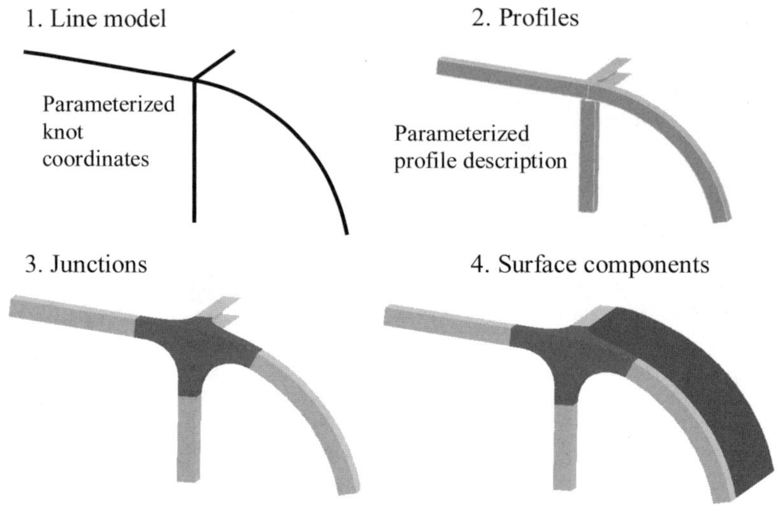

Fig. 2 Line-model based design.

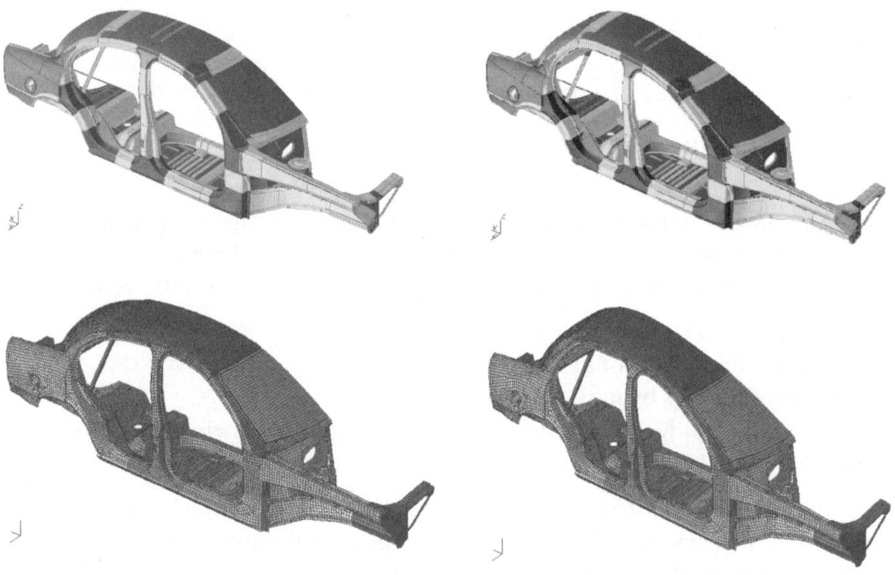

Fig. 3 Example for changes of the model shape in SFE CONCEPT® and in the corresponding automatically generated FE-mesh (Zimmer and Schumacher 2008).

4. Process chain with integration of processing dependent material parameters

The material of cast components is typically not homogenous. Due to the cast process and the solidification conditions, material properties and cast defects depend on the location in the component. Such inhomogeneities influence the properties of a component but also the optimal shape after shape optimization. The following example illustrates the influence of the cast porosity on the failure behaviour and the optimal shape of a flat tensile test specimen. In the optimization the probability of having the higher stresses outside of the measurement area should be minimized. Therefore, the optimization problem is defined as follows:

$$\min\left[\Phi\left(\frac{\mathrm{E(f)}}{\sqrt{\mathrm{Var(f)}}}\right)\right] \quad \text{with } f = \sigma_{max,2} - \sigma_{max,1} \tag{1}$$

Where σ_{max1} is the maximal stress in the measurement area, σ_{max2} the maximal stress in the fillet, *E(f)* the expectation value of the objective function *f*, *Var(f)* the variance of *f* and Φ the cumulative distribution function of the standardized normal distribution. Assuming homogenous material, the shape optimization results in a reduction of "f "but there exists no shape with maximal stress in the measurement area. For evaluating the influence of pores on the failure probability in the measurement area a defined number of elements in the FE-model of the test specimen has been deleted by means of a random generator. Based on 500 samples the values μ and σ of the standardized normal distribution were calculated to get the corresponding probability density function and the cumulative distribution function. It can be seen that there is an area under the curve, where the objective function f is negative. That means that the specimen will fail in the measurement area. Figure 4 displays the result for 1% porosity.

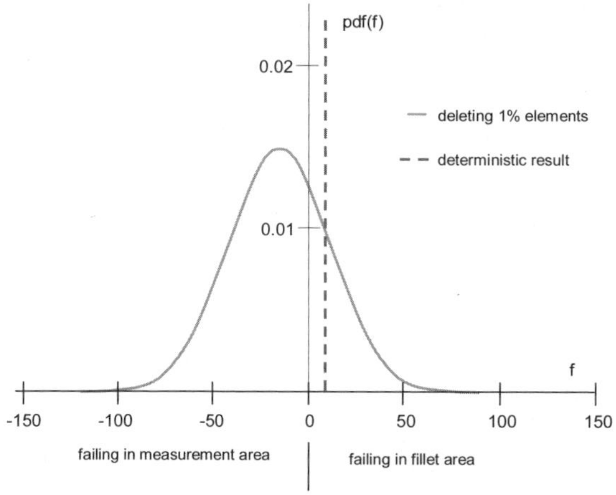

Fig. 4 Influence of inhomogenously distributed pores on the probability density function pdf(*f*).

For different porosities a shape optimization has been performed (Schumacher *et al.* 2009). The optimal shape of the fillet was described by a hyper-elliptical curve, which depends strictly on the porosity. By integrating the cast simulation into an automated CAD/CAE process chain, the here randomly generated pore distributions can be weighted.

5. Conclusions

Different processes in the development of components can be linked automatically in process chains in "batch"-mode. Integration of the CAD-system is of central importance. The CAD-program is often used at the start of the process chain. However, if conceptual changes are possible in the process chain, the CAD-program may be used several times during the automatic process. The presented schemes are implemented and controlled by the user without involving providers of CAD-software.

Acknowledgments Part of the research was funded by the German Ministry for Education and Science (BMBF, No. WING 03 X0509).

References

NOESIS Solutions N.V. (2008): *OPTIMUS 5.3 SP1 Users Manual*. Leuven, Belgium

Schumacher, A. (2005): *Optimierung mechanischer Strukturen – Grundlagen und industrielle Anwendungen*, Springer, Berlin Heidelberg

Zimmer, H.; Schumacher, A. (2008): "*SFE CONCEPT: Funktionsauslegung und Optimierung in der frühen Entwurfsphase*", in: Tecklenburg, G. (Hrsg.): Die digitale Produktentwicklung, Expert-Verlag, Renningen

Schumacher, A., Hakim, K., Bartsch, M. (2009): "*Shape optimization of cast parts with porous material*", Proc. of the 8th World Congress on Structural and Multidisciplinary Optimization, June 1-5, 2009, Lisbon, Portugal

TRANSITIONING TO THE FUTURE OF HYDROGEN MOBILITY

J. I. Galindo

President of Alset Technology LLC,
Bouchard 599 Capital Federal. Buenos Aires, Argentina
E-mail: jigalindo@alsetglobal.com

Abstract Hydrogen is now being accepted as the most sustainable and efficient alternative fuel for vehicles, and most car manufacturers have focused their hydrogen-related research and development on fuel cell technology. But, even though these types of vehicles have achieved great levels of efficiency by combining the best of both worlds (the most efficient type of propulsion "electrical motor" and the most efficient energy storage method "hydrogen"), the sale of this type of vehicle, unfortunately, will not be fully manifested without a pervasive hydrogen-refueling network. The transition to the hydrogen economy won't happen overnight. The solution is a technology that permits a gradual transition from fossil fuel to hydrogen that provides an acceptable autonomous range for hydrogen-based mobility, that is economically feasible, and that can guarantee sustained independence from foreign energy sources. After four years of intensive Research and Development in the hydrogen field, Alset has developed an advanced transitional hydrogen engine technology for utilizing a bi-fuel (hydrogen and gasoline) cross platform for internal combustion engines (ICE). This technology can be integrated in car manufacturers (OEM's) series production lines and, in addition, can for retrofitted into existing gasoline vehicles allowing them to run bi-fuel as well. Through high-level simulations and rigorous testing procedures the Company have demonstrated the ability of the Alset bi-fuel cross platform to promote a safe, efficient and clean way to power mobility, while allowing the manifestation of massive consumer demand for hydrogen in the marketplace.

1. Introduction

As the global population continues to grow and countries strive to improve their standard of living, energy use will continue to grow significantly resulting in more competition for limited resources and an increase in pollution. This is pushing world leaders and citizens alike to demand new and sustainable economic growth based on renewable energy, resource efficiency and clean air.

This has created a global race to build the car of the future that will address these concerns. Researchers from different parts of the planet are tackling the

J. Wellnitz et al. (eds.), *Sustainable Automotive Technologies 2010*,
DOI 10.1007/978-3-642-10798-6_14,

challenge of creating automobiles that will run on cleaner and renewable energy sources.

We are in the beginning of a major transformation away from carbon-based energy sources to new fuels and technologies. Innovation will be the driving force on the redefining of two key industries: *Energy* and *Transportation.*

By incorporating technologies that allow the use of clean fuels in the transportation sector, the automobile industry will be the leading industry to create the demand for new clean fuels.

The cleanest, most sustainable and practical of these fuels is hydrogen.

2. Hydrogen as an energy carrier

First of all, it should be noted that hydrogen is not an alternative for producing energy but for storing it. Today, energy storage is accomplished mainly by the use of liquid hydrocarbons (such as gasoline).

Fossil fuels, however, are destined to disappear, which means that not only a new energy source must be developed, but also a new method for storing energy.

This is where hydrogen distinguishes itself.

A key measure for determining the best replacement fuel is its energy density. This is especially important when the storage needs to be mobile (as in a car, aircraft or ship).

Hydrogen has the highest energy density per unit of weight of all fuels known, including ethanol, gasoline, bio-diesel, diesel fuel, natural gas, etc. This means that for a given storage weight, hydrogen holds the largest amount of energy. The technology for storing and using hydrogen is at an advanced stage of development, although some final developments are still necessary, especially regarding hydrogen's volumetric density. However, rapid progress is being made in solving these final barriers to hydrogen's full viability as a functional energy source.

Secondly, hydrogen is a clean fuel. Whether it generates energy by combustion or by fuel cells, its emissions are carbon dioxide and carbon monoxide free. Also, Hydrogen can be produced from water, which allows a renewable and geographically even-distributed source of this element. This will dramatically reduce the geopolitical and national security concerns currently associated with fossil fuels.

Finally, another advantage of hydrogen is related to its use in internal combustion engines, where the efficient mobility obtained allows Hydrogen to position itself as a crucial fuel regarding the market share it can gain during the next years compared to several alternatives.

3. Transition technology

Despite the major technological advances in storage methods, production and distribution, the hydrogen economy has been on hold for over a decade. Changing to a hydrogen fuel based economy and culture, and bringing fuel cells or other hydrogen technologies to the market remains a challenge for the industry. Current projections estimate a fourteen year window before a true return on investment given the various stages of development and implementation. This represents a barrier preventing the hydrogen vehicle industry from manifesting. It is highly unlikely that any private company will invest in a commercial hydrogen venture unless it believes that it will be able to compete against existing fuels and turn a profit in a reasonable timeframe.

Therefore, there is a need for a transition period, rather than an abrupt change, where both conventional fuels and hydrogen are available and used in conjunction in the same vehicle and in the marketplace giving flexibility to end users as they gravitate into this new technology.

The solution to meet this transitional requirement is the manufacturing of bi-fuel vehicles, running on both hydrogen and gasoline using current internal combustion engine technologies. This bi-fuel approach will stimulate the creation of a hydrogen refueling network thus allowing for a full transition to a hydrogen powered vehicle economy.

Two car manufactures, also referred to as Original Equipment Manufacturers or OEM's, have started down this path. BMW and Mazda each have produced bi-fuel (hydrogen and gasoline) vehicles, however their technology is limited to new vehicles and a specific make and model thus not having any cross application to other brands and other type of vehicles and engines.

3.1 Alset bi-fuel technology – a cross platform

For the past four years, Alset Technology LLC has been working on developing the technology and the strategy to bridge the gap towards a sustainable hydrogen economy. In alliance with FVT-TUG, the Austrian research center in internal combustion of hydrogen, Alset developed the first fully engineered bi-fuel (*hydrogen and gasoline*) cross platform to retrofit existing gasoline vehicles to run bi-fuel. This same bi-fuel cross platform could be integrated in car manufacturers' (OEM's) series production lines of new vehicles with a slight reconfiguration.

This bi-fuel concept consists of one engine with two fuel systems on board (hydrogen and gasoline) giving flexibility to end users as they gravitate into this new technology. This ensures that if a driver runs out of hydrogen and no hydrogen refueling station is at hand, he/she can continue driving with gasoline. This will provide the ultimate freedom required for hydrogen end users. The Company carried out a complete engineering development of the system comprising strict safety requirements, hydrogen storage system, thermodynamic processes, emission targets, components, materials, electronics, and all required applications on an engine to run with hydrogen.

This bi-fuel approach will stimulate the creation of a hydrogen-refueling network thus allowing for a full transition to a hydrogen powered vehicle economy.

This is a key enabling technology to manifest massive consumer demand for hydrogen.

3.2 Ongoing mobility with either hydrogen or gasoline

Cars with the Alset bi-fuel system will increase their autonomy range by the use of hydrogen. It is estimated that a vehicle with bi-fuel operation will be able to cover a distance of approximately 125-185 miles (200-300 kilometers) on hydrogen and up to 310 miles (500 kilometers) on gasoline. This produces a great autonomy when compared to alternative vehicles, specially the battery-related cars, where they present several autonomy limitations.

3.3 Enhanced efficiency of hydrogen.

Hydrogen has excellent qualities and properties that give the combustion engine a very high standard of efficiency, greatly exceeding the efficiency of current gasoline engines. Additionally, and importantly, according to the German Technical Inspection Authority (TÜV), running a car on hydrogen can be just as safe as running a vehicle on gasoline.

3.4 Technical findings in the development of bi-fuel prototype

The Alset bi-fuel prototype presented excellent levels of engine performance. Table 1 shows the specific power output of the Alset Bi-fuel prototype when compared to other currently available hydrogen vehicles in the market.

Table 1 Comparison between several vehicles on the market and Alset's H2, Bi-fuel prototype

Model	Engine	Capacity	Power	Torque	Specific Power
BMW Hydrogen 7	V12 bivalent	6.0 lt.	260 HP	390 Nm (278 lbft)	43.5 HP/lt.
Ford Focus C-Max	4cyl – inline monovalent	2.2 lt.	112 HP	Unknown	49.8 HP/lt.
Quantum Prius H2-Hybrid	4cyl – inline hybrid	1.5 lt.	71 HP	111 Nm (79.3 lbft)	47.4 HP/lt.
Ford Shuttle E-450	V10 monovalent	6.8 lt.	272 HP	1110 Nm (786 lbft)	40.0 HP/lt.
Alset H2 Bi.Fuel 1.0	4cyl – inline bivalent	2.0 lt.	150 HP	390 Nm (278 lbft)	75.2 HP/lt.

4. Conclusions

Without any doubt sustainable growth is the biggest challenge facing humanity today. We are at the beginning of a major transformation away from carbon-based energy sources to new fuels and technologies. Innovation will be the driving force while redefinition of two key industries takes place: Energy and Transportation. Nevertheless, there is a need for a transition period, rather than an abrupt change, where both conventional fuels and hydrogen are available and used in conjunction in the same vehicle and in the marketplace giving flexibility to end users as they gravitate into this new technology. The solution to meet this transitional requirement is the manufacturing of bi-fuel vehicles, running independently on hydrogen or gasoline using current internal combustion engine technologies. This bi-fuel approach will stimulate the creation of a hydrogen-refueling network thus allowing for a full transition to a hydrogen powered vehicle economy. This is a key enabling technology to manifest massive consumer demand for hydrogen in the marketplace.

GREEN CAR TECHNOLOGIES – CHALLENGES AND OPPORTUNITIES

A. Subic, L. Koopmans, M. Leary

RMIT University, Bundoora, Victoria 3083, Australia
E-mail: Aleksandar.Subic@rmit.edu.au

Abstract Climate change and the declining supply of conventional fuels due to peak oil have rapidly increased the need for new sustainable vehicle platforms based on fundamentally different technologies, including environmentally friendly alternatives to conventional engines, lightweight structures, use of recyclable and renewable materials, energy efficient manufacturing etc. Future vehicles have to be green, that is have extremely low energy consumption and emissions, in order to meet the challenges posed by global warming, population growth, finite fossil fuels and air quality. Which technologies will be the most effective from the life-cycle perspective will be explored in this paper, and key challenges and opportunities identified.

1. Introduction

In recent years car sales dropped dramatically primarily because of the global financial crisis and increase in fuel cost. Although the global financial crisis poses a great challenge to the car-industry, there lies an even bigger challenge ahead and perhaps one of more permanent nature. Future vehicles will need to be "green", that is have extremely low energy consumption and emissions, in order to meet the following global challenges:

- global warming
- population growth
- finite fossil energy supply
- air quality.

In order to produce green vehicles, car-manufacturers must focus on the following main areas of concern:

- Reduce CO2 emissions and hence green house gas (GHG) emissions
- Increase efficiency and hence reduce consumption of energy and natural resources
- Reduce emissions and hence improve air quality
- Reduce the overpopulation of vehicles on the roads.

J. Wellnitz et al. (eds.), *Sustainable Automotive Technologies 2010*,
DOI 10.1007/978-3-642-10798-6_15,

While car manufactures generally accept that they need to invest in new technologies to meet the objectives experience to date shows that many manufacturers are reluctant to make radical changes unless forced by legislations or consumer demand. Although the automotive industry sector is spending more on research than any other private sector in Europe (e.g. €20 billion = 4% of turn-over) the average CO_2 emissions were around 154 g/km in 2008. This reduction was mainly caused by the increased market share of diesel powered vehicles which increased from a little over 20% of the market-share in 1995 to around 50% in 2008. It was not until the European Commission in January 2007 proposed to legislate CO_2 emissions in 2012 that new more environmentally friendly technologies started to emerge in order to radically reduce CO_2 emissions for spark ignited powered vehicles. There are some exceptions to this practice. For example, Toyota launched their Hybrid Prius in 2000 as part of their marketing strategy for a more greener and innovative image and accomplished 120 g/km CO_2 on the EC00 European driving cycle. Also, BMW reached 140 g/km by introducing a 2l, 4cyl. gasoline engine with stratified combustion technology in their 118i model (Subic and Koopmans, 2010).

It is difficult to predict how the development of green car technologies will evolve beyond the legislative targets. In the current global economic context many car manufactures have great difficulty to invest into long term research and development and are primarily oriented towards shorter term gains based on "low hanging fruits". Also, having realised that there is no single technological solution to the problems of climate change and peak oil, we can expect a more rapid introduction of a range of alternative green car technologies in the following years. Which technologies will be more effective from the lifecycle perspective, and what are the main technological challenges and opportunities facing the global automotive industry and research in general? This paper will explore the potential answers to these questions.

2. Challenges

As outlined in the previous section, the paper will explore the following main challenges and their implications on green car technology development:

- global warming
- population growth
- finite fossil fuel supply
- air quality.

2.1 Global warming

There is ample evidence available today that global warming is occurring. The most significant amount of green house gases (GHG's) is produced by the burning of fossil fuels (70.9%), which include carbon dioxide (CO2) and methane (CH4). The concentrations of these GHG's have by far exceeded the natural range over the last 650000 years. While there are varied estimates of how much road vehicles (passenger cars and light-commercial vehicles) contribute to overall GHG's emissions, recent studies estimate this to be around 16% (Brown, 2009). It seems that greenhouse gas emissions from road transport are rising. In its review of the of the European Community strategy to reduce CO2 emissions from passenger cars and light-commercial vehicles published in April 2007, the European Commission wrote "the transport sector is one of the only sectors whose emissions keep increasing, making it harder for the EU to meet its Kyoto commitments by jeopardising the progress made by other sectors."

2.2 Population growth

Already there are more than 900,000 vehicles on the roads world wide generating high demand on fossil fuel supply. With the increase in world population the growth in the number of vehicles on the roads will continue. Hence the energy consumption will increase as well. Countries like India and China in particular will increase their energy consumption dramatically partly due to the modernisation of transportation and the growing demand for personal vehicle ownership due to increased income per capita (EIA, 2009). Development of smaller and cheaper vehicles for these markets (such as Tata Nano) will further fuel this growth.

2.3 Finite fossil fuel supply

Predicting whether there will be enough oil in the future to meet the world demand is extremely difficult, especially since oil producing countries are unwilling to reveal or do not have the knowledge of how much oil reserves they have left. Bentley and Boyle (2008) collected a range of forecasts regarding conventional oil production. They compared many scenarios and methodologies and came to the conclusion that an oil peak is to be expected before 2020. However the rate of decline after the peak is expected to be reduced due to the discovery and utilization of un-conventional oils. In any case, it can be concluded with some certainty that fossil fuel supplies will not be sufficient to meet the growing demand in the future.

2.4 Air quality

Transportation is one of the main sources of four out of six identified common air pollutants which are a threat to common health, the environment and property:

- Ozone
- NOx
- CO
- PM (particulate matter).

A number of remedial technologies have been developed to deal with this problem resulting in advanced exhaust after-treatment systems. For gasoline engines, the three way catalyst with lambda sensor and lambda 1 engine operation was developed, having the potential to reduce the polluting exhaust emissions down to almost zero. For diesel engines expensive and complex systems are being developed to cope with the tough emissions requirements. Diesel emissions after-treatment systems containing oxidation catalysts, NO_x after-treatment parts and particulate filters can count up to 30% of the total engine price. Emissions standards and regulations are driving further improvements. Unfortunately, the requirements relating to exhaust emissions have a negative impact on fuel consumption of both diesel fuelled compression ignition (CI) engines and gasoline or gas fuelled spark ignition (SI) engines. Regeneration strategies for the particulate filter and NOx after-treatment system, combustion strategies with inefficient phasing of the combustion, and increased back-pressure in the exhaust system all increase CO_2 emissions.

3 Opportunities

The internal combustion engine proved to be very successful to date because it utilizes a portable energy source (fuel), which has a large energy density, at least compared to the other portable energy sources that are available today (e.g. batteries). It is most likely that the internal combustion engines will be part of our transport systems for many decades to come as it is not technically feasible (nor sustainable) to replace this technology rapidly. However, because of being the polluter and contributor to the challenges described in previous section, the internal combustion engine and associated technologies must improve their efficiency. Therefore, many opportunities for green car technologies lie in the improvement of efficiency of the spark-ignited engine.

Figure 1 below shows the breakdown of the energy flow from tank-to-wheel by Volvo Car Corporation. Only 6% of the energy input is utilized in inertia while driving the European driving cycle for emissions and fuel consumption assessment with an EU C/D vehicle.

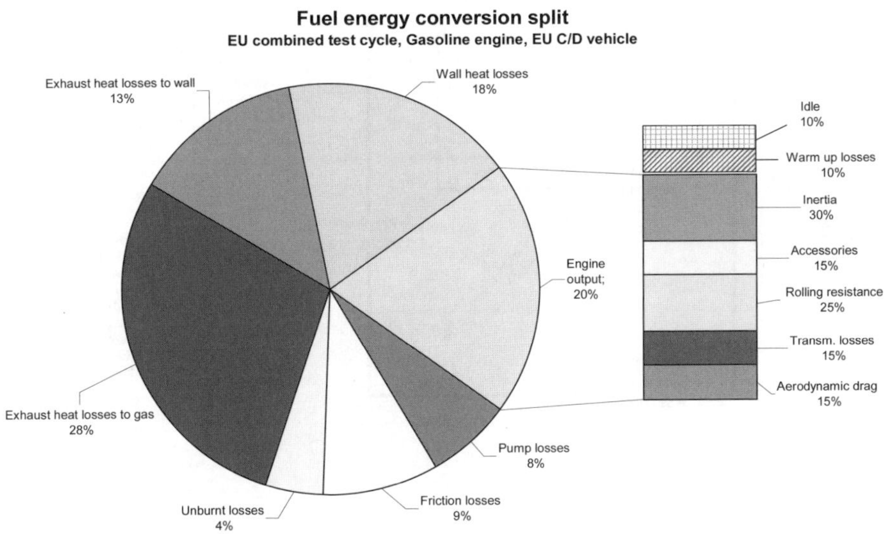

Fig. 1 Fuel energy conversion breakdown for an EU C/D vehicle with a gasoline engine. *Courtesy Volvo Car Corporation.*

The fuel conversion split can be divided into vehicle related energy losses and engine related energy losses. Opportunities for the vehicle lie within weight reduction and aerodynamic drag reduction. Also, there is much gain to be achieved from matching the transmission and engine in such a way so that it operates more in its most efficient region during normal driving. The idle losses can be reduced by utilizing start/stop technology by using for example an integrated starter generator (ISG) which can also regain some of the inertia energy by regenerative braking. Opportunities for improving the gasoline engine lie in the reduction of pumping losses, thermal and mechanical losses, and by improving the combustion efficiency. However, the maximum efficiency feasible is limited by the operating principles of the SI engine namely the Otto cycle.

The majority of future SI engines will be small displacement volume engines, turbocharged and/or supercharged (like the diesel engines). Replacing a larger naturally aspirated engine with a small boosted engine so that the engine operates more frequently closer to the maximum efficiency region (reducing pumping losses) is called downsizing. Figure 2 shows how supercharging technology is positioned with respect to engine downsizing and associated technologies.

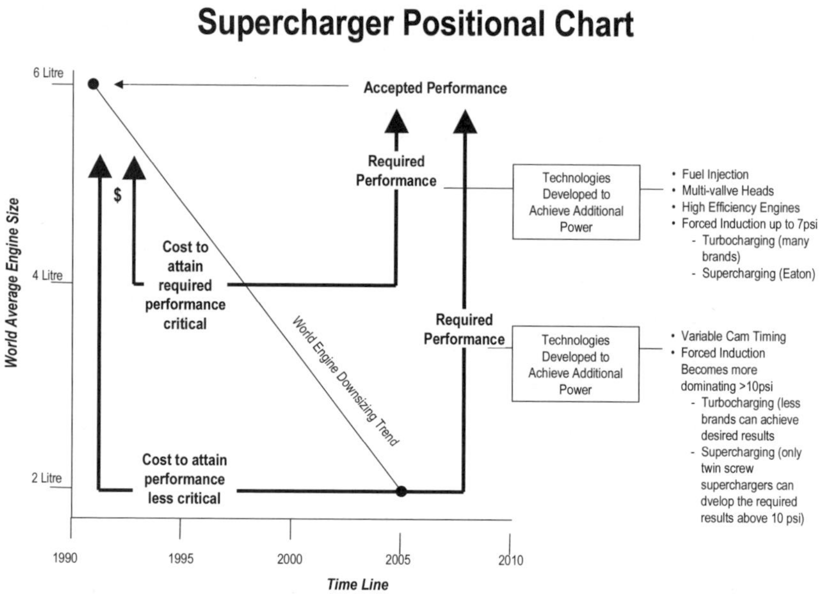

Fig. 2 Engine downsizing and supercharger technology.

In addition to downsizing, future engines will also be increasingly equipped with direct fuel injection. Injecting the fuel charge directly into the combustion chamber allows the heat of evaporation to be drawn from the inducted air instead of the manifold, therefore cooling down the charge and decreasing the knock sensitivity of the engine. As a result, the compression ratio can be increased improving the thermal efficiency of the engine. With special design of intake ports, combustion chamber, piston, sparkplug and injector location, stratified combustion can be made possible. Operating the engine with stratified combustion (like the diesel) reduces both pumping losses and wall heat losses. Unfortunately this technique is limited to only part load and extra NOx and perhaps particulate mater exhaust emissions after-treatment systems have to be added in order to meet the stringent emissions legislation levels.

Other new vehicle technology trends include (Subic and Koopmans, 2010):

- Alternative valve trains (pumping efficiency and thermal efficiency through Atkinson cycle)
- Thermal management (wall heat losses and friction)
- Improved bearing technology and materials (friction and wall heat losses)
- Homogeneous charge compression ignition (pumping losses, thermal efficiency and wall heat losses)
- Variable displacement volume (pumping losses)
- Exhaust heat recovery.

Alternative fuels other than the conventional fuels like gasoline and diesel also provide significant opportunities in the future. These fuels can be either fossil fuels like natural gas, propane and butane or bio fuels like bio-gas, bio-alcohols, hydrogen or vegetable/plant oil. In order to lower CO_2 emissions and improve sustainability, bio-fuels are preferred because of the natural cycle (plants consume CO_2 while growing). The disadvantage is the energy consumption while producing bio-fuels suitable for transport which can be a multiple factor of the energy released while burning (depending on the source). Another disadvantage is the maximum production rate of bio-fuels which is only a fraction of the demand for transport. But, for economical and environmental reasons many governments now require a 5 to 10% bio-fuel amount blended with the conventional fuel.

Dedicated alternative fuel vehicles have a low rate of success at present because of the lack of infrastructure. This has been one of the impediments to the introduction of hydrogen vehicles or plug-in electric vehicles. Therefore, the development of engines that can operate on both conventional and alternative fuels preferably involving more than two options may provide significant benefits in such circumstances. The real opportunity can be found in the development of new technologies that combine gaseous fuels and liquid fuels in direct injected, turbo-charged engines not to mention stratified spark-ignited engines with one fuel-injector.

4 Conclusion

The paper presented in a systematic manner some of the main sustainability challenges facing the automotive industry sector. A comprehensive discussion of the challenges and opportunities has been provided, which was used as a platform to identify potential green car technology development opportunities.

References

Subic A. and Koopmans L. (2010). Global Green Car Learning Clusters. *International Journal of Vehicle Design*, Vol. 53, No. 1 (in press).

Brown S. (2009). *Vehicle Manufacturers in the Race for Environmental Success*. Research Report, ABOUT Automotive, UK.

Bentley R. and Boyle G. (2008). Global Oil Production: Forecasts and Methodologies *Environment and Planning B: Planning and Design*, Vol. 35, pp 609 - 626.

A NEW HYBRID MATERIAL FOR INCREASED IMPACT TOLERANCE

U. Burger[1], J. Wellnitz[2]

[1]MBDA Deutschland, Landshuter Str. 26, 85716 Unterschleißheim, Germany, E-mail: uli.burger@mbda-systems.de

[2]Hochschule Ingolstadt, Esplanade 10, 80549 Ingolstadt,Germany E-Mail: joerg.wellnitz@fh-ingolstadt.de

Abstract The scope of this work is the investigation of the impact behavior of fibre composites reinforced with metallic chain mail, especially the field of high velocity impact. Besides the first production of samples of the new material combination experimental investigations are carried out on coupon testing level. These include 3-point-bending tests, charpy-impact tests and ice impact tests at high velocity on a self-developed gas gun. Testing is accompanied by looking at the possibilities of numerical simulation of the new material and the impact event with the help of the finite element method (FEM). Based on the experiments and simulations first material values and characteristics could be obtained. For reference, non reinforced carbon fibre reinforced plastics (CFRP) are investigated additionally.

1. Introduction

The scope of this analysis was the basic investigation of a new hybrid composite material. It consists of CFRP which is enhanced by metallic chainmail in order to increase the impact tolerance.

For comparative reasons also pure CFRP specimens has been investigated. Two CFRP configurations have been used for the first specimens. One was unidirectional AS4-graphite-fibre with PEEK-matrix. This material is widely used in the airplane and aerospace industry and certified for these applications. Due to cost aspects also a common, non-certified prepreg of carbonfibers and epoxy resin was used for specimen production. In Figure 1 the principle lay-up of the hybrid composite is shown.

J. Wellnitz et al. (eds.), *Sustainable Automotive Technologies 2010*,
DOI 10.1007/978-3-642-10798-6_16,

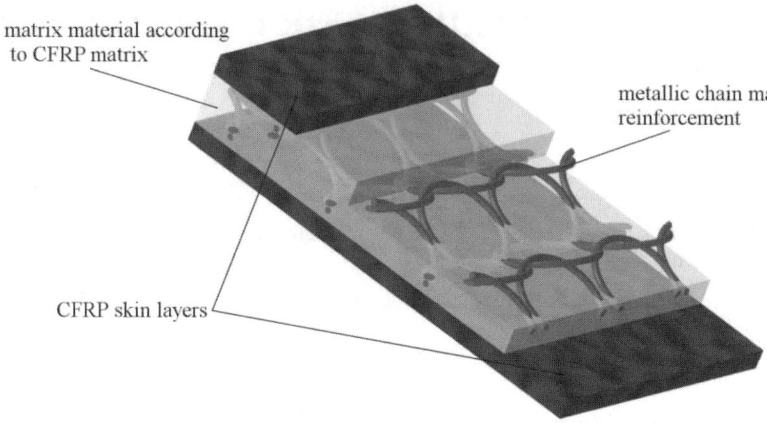

Fig. 1 Principle layup of CarbonICE.

2. Experimental work

Extensive experimental work has been done during the first evaluation phase. The following types of specimen have been tested:

- Type A: Sigratex Prepreg CE 8201-200-45S symmetric skin lay-up with a core of chainmail Titan Grade 2
- Type B: identical fiber lay-up as type A, without chainmail
- Type C: same thickness as type A, the chainmail core is replaced by additional fiber layers
- Type D: APC-2/AS4 symmetrical skin lay-up with a core of chainmail steel 1.4306
- Type E: identical lay-up as type D without chainmail

2.1 Charpy-impact tests

In order to investigate the impact strength of the composite of chainmail and CFRP impact tests according to DIN EN ISO 179 has been performed. The derived value out of these tests is the Charpy-Impact-Strength α.

Impact-strength herein is the energy absorbed by the specimen when broke through during the test related to the initial section. In order to be able to assess the effectivity of the chainmail enhancement also reference specimen as described before has been tested.

With the evaluation of the test and the identification of mechanical properties first statements upon the behavior of the material at punctual, dynamical loads could be drawn.

Comparing the different types of specimen the following ascertainments could be made:

- In contrast to the pure CFRP-types the CarbonICE-specimens weren't completely chamfered. Additionally high residual strength could be obtained. The CFRP-layers are backed by the chainmail-core and a complete failure at the side of impact could be avoided. Though the epoxy resin around the metallic rings is broken at the impact region no ongoing failure could be seen. Without a stiff surrounding matrix the chainmail rings are now able to move and dissipate parts of the impact energy. Some of the rings in the impact section are slightly plastically deformed.
- As announced before a large deformation way for the chainmail is now possible by breaking of the matrix. Thus the upper CFRP layers have kind of elastic bedding. This prevents the failure of the upper surface layers. As described here only delamination and single fiber failure in single areas are obtained and not over the total width and thickness of the specimen.

These increased characteristics could also be seen by the impact strength values of the individual types, as figure 2 indicates. Thus an increase can be stated of around 49,5% compared to type B and an increase around 63% compared to type C.

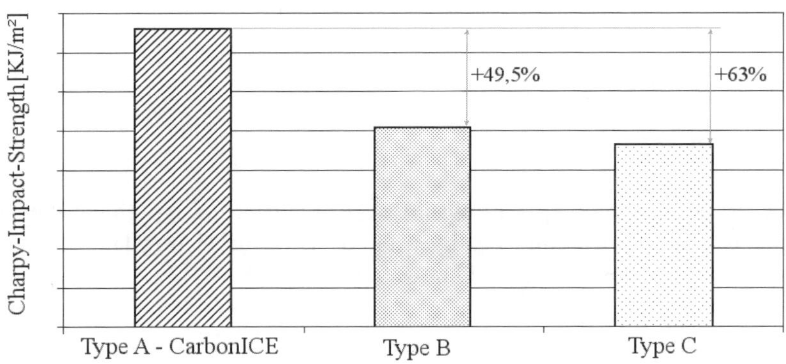

Fig.2 Comparison of Charyp-Impact-Strength for types A, B and C.

A comparison of types D and E shows the same results, nevertheless the increase is not as high as before. This could be explained by the already improved impact strength of the CFRP material (AS4-fibers, PEEK-matrix) used for these types. But here also an increase of around 11% could be achieved by using a chainmail layer.

An increase compared with issue index E is mainly due to the intermediate layer of PEEK and chainmail. The impact energy could be additionally dissipated by the deformation of this layer. The chain mail prevents a complete breakage of the matrix and therefore the upper CFRP layers are backed. By this the complete breakage of the specimen could be avoided.

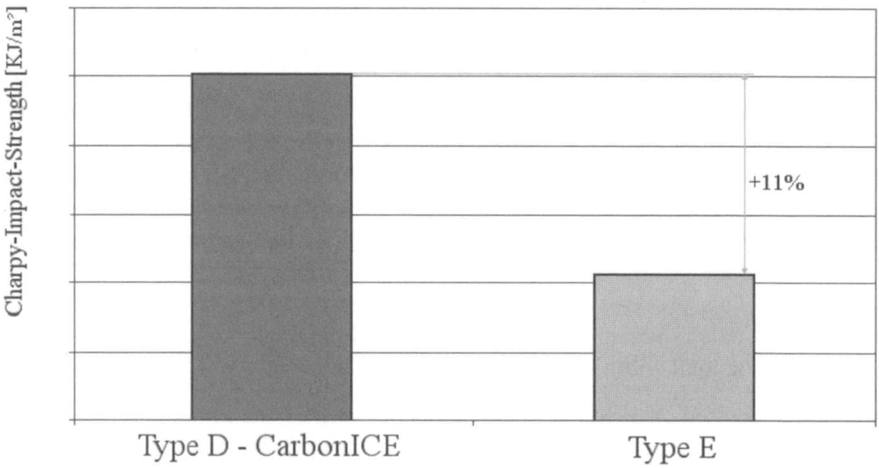

Fig. 3 Comparison of Charpy-Impact-Strength for types D and E.

2.2 *Ice impact tests*

In order to perform high speed impact tests of realistic foreign object scenarios a gas gun was developed followed by a build up in the testing laboratories. The principle set up of the test rig could be seen in Figure 4.

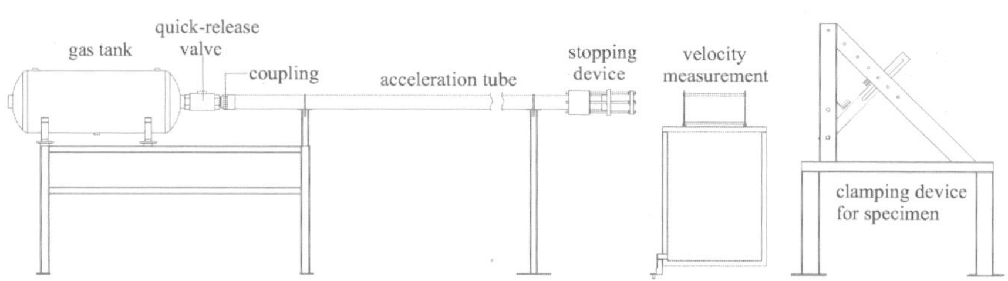

Fig. 4 Principle set up for ice impact tests.

To obtain detailed information during the impact high speed camera movies has been taken.

Rectangular specimen plates with the same types of layups as described above have been tested. As a projectile a simulated hail ice sphere (SHI) was used. The weights of the spheres were around 50g with an impact velocity of 140 m/s.

In figure 5 the backside of the impacted plates of types A, B and C are compared.

Fig. 5 Backside of impacted specimen types A, B and C.

By looking at the apparent damage of the different types tested the following statements could be made:

- The CarbonICE-specimen, type A, shows less damage than the other two specimen types, B and C. This could be seen by looking at the smaller damage area.
- However, also for the CarbonICE-specimen a complete failure in through thickness direction could be obtained. In the direct environment of the impact the chainmail is broken. The breakup of the rings took place at the welding point and leads therefore to the fact that the quality of the welding spot was not sufficient enough. In the following the chainmail layer was not able to back the fiber layers and a complete failure of the structure appeared.
- Also delamination occurred at the impact section between the skin fiber layers and the chainmail core. Nevertheless, the same behavior could be detected for the other types B and C. Here wide rupture in through thickness direction appears. As a result a breakup of the structure followed and contrary to type A parts of the specimen broke away for types B and C.

For specimen types D and E no penetration of the specimen took place. The appearing damage, primary at the rear side of the impact, could be seen in figure 6.

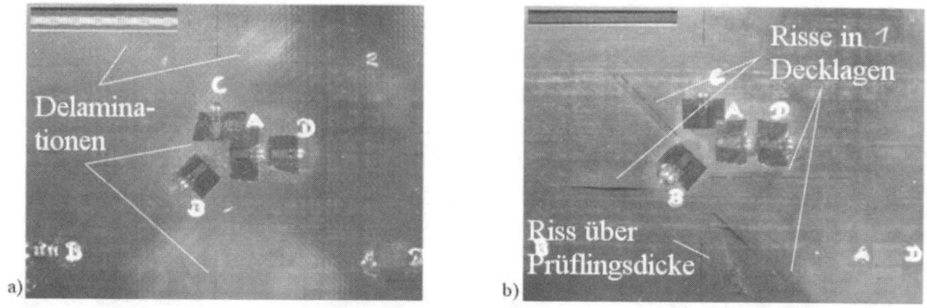

Fig. 6 Backside of impacted specimen types D and E.

Considering the above pictures the following analyses could be drawn:

- On the rear side of the unreinforced specimen, type E, cracks occur. These cracks are star shaped leading away from the point of impact. In the direct environment of the crack also delaminations between the CFRP layers occur.
- The vertical crack goes beyond the upper fiber layers and leads to a complete through thickness failure. So, it could be easily detected from both sides of the specimen as shown in Figure 6 b.
- The chainmail enforced specimen, type D, shows no visible damage at the surfaces. Wether on the front nor at the rear side cracks or similar damage could be obtained, as could be seen in Figure 6 a.
- At the rear side of the specimen though delaminations could be obtained. It could be assumed that this took place between the chain mail core and the CFRP skin. Directly at the point of impact no delaminations could be detected. But here a plastic deformation could be seen with a depth of around 4mm and a hoper shape.

3. Simulation

Together with the experimental efforts finite element simulations has been performed in order to evaluate the possibilites to investigate the new material. The main effort hereby lied in the simulation of ice impact as tested before. By using the Smooth Particle Hydrodynamics Method to simulate the hail ice sphere and common Langragian modeling techniques for the hybrid material very good results could be obtained.

Based on these first results the modeling of the hybrid material was enforced and excellent correlation between the above shown test results and the simulation was achieved as shown in Figure 8.

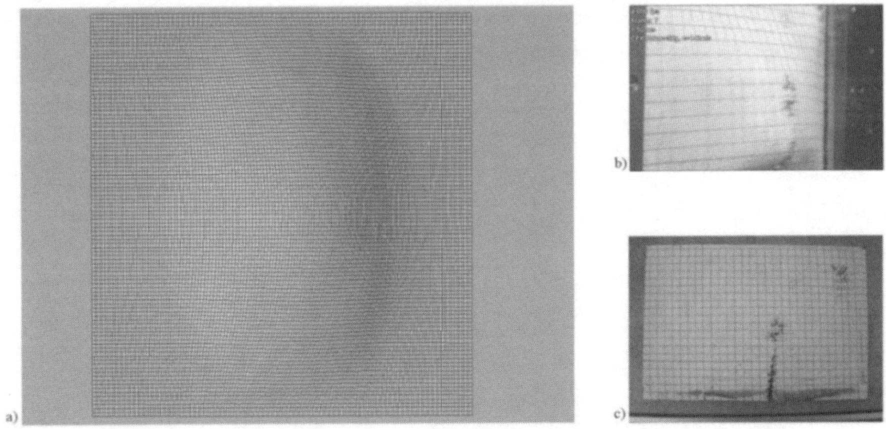

Fig. 8 Comparison of simulated ice impact on specimen type A - CarbonICE and experiment

This progress enables the further investigation of the material in order to optimize the special characteristics towards different impact scenarios and beyond.

4. Conclusion

By looking at the results of the first evaluation phase the enormous potential of the new hybrid material containing common CFRP structures and chain mail enforcement is clearly visible.

The impact strength and tolerance could be increased by an essential amount depending on the based CFRP structure. This effect could also be transferred to other FRP structures so that a broad band of application could be imagined in various fields wherever needed.

References

Burger U. (2009) Impactverhalten von hybriden Verbundwerkstoffen mit metallischem Ringgeflecht. PhD thesis. Technische Universität Berlin, Germany.

THE IMPORTANCE OF HYDROGEN AS SECONDARY ENERGY CARRIER FOR RENEWABLE PRIMARY ENERGIES

J. Toepler[1], R. Wurster[2]

[1]German Hydrogen and Fuel Cells Association, DWV, Berlin, Germany
[2]DWV and Ludwig–Bölkow-Systemtechnik, LBST, Munich,Germany

Abstract This study addresses the future energy situation with the expected reduction of fossil primary energy supply and increasing contributions by renewable energies within the next decades. The potential of renewable energies to cover the energy demand is described as well as cost reductions of wind power and photovoltaic energy. The increasing use of renewable vehicle fuels is investigated with special focus on hydrogen in short and long term considerations. As a result the study will show an energy supply gap between decreasing fossil energies and increasing renewable energies which can only be bridged by a more efficient general use of energy. Hydrogen will be proven as best synthetic fuel by "well-to-wheel analyses" both with respect to energy efficiency, to environmental analysis as well as to customer satisfaction.

1. Introduction

Hydrogen as a universal energy carrier can be produced as secondary energy from a wide variety of primary energy sources. Considering the whole energy chain from production to end-use, hydrogen, used in fuel cells to power transport and stationary applications, can provide significant benefits in terms of greenhouse gas emissions and local pollutants through increased efficiency and/or lower rate of emission per end-use energy unit.

There is a broad consensus that the future powertrain is electric. Battery vehicles would be the most energy efficient solution. However, by evaluating different secondary energies with respect to energy density, efficiency, environmental aspects and subject to customer performance expectations and infrastructure development (both electricity grid and hydrogen infrastructure), hydrogen in combination with fuel cell vehicles have excellent perspectives already in the medium term but definitely in the long term.

J. Wellnitz et al. (eds.), *Sustainable Automotive Technologies 2010*,
DOI 10.1007/978-3-642-10798-6_17,

2. Future Energy Supply

The aggregate supply of fossil and nuclear energy is expected to peak by around 2015.

After the peaking of the oil supply (today) [1], the global supply of natural gas, coal [2] and nuclear fuel [3] is expected to reach a combined maximum by around 2020 – at the latest. This will have significant impacts on the total energy supply. With peak oil we are entering into a transition phase towards a post-fossil energy area.

The limitations in availability of fossil energy resources as well as the threat of climate change to biosphere have led to the formulation of political goals with regard to security of energy supply and especially reduction of greenhouse gas emissions. All the underlying issues can be addressed in an efficient and sustainable way by energy conservation, by the increased use of renewable energy sources and by the use of hydrogen and fuel cells.

In the long term, renewable energies will be able to provide more energy than all fossil and nuclear fuels will have been able to supply at their aggregate peak. On a global level, solar energy and wind power can become the major pillars of our energy system. (Figure 1)

The transition from fossil fuels to renewable energies is a transition from primary energy fuels (i.e. crude oil, natural gas, coal) to electricity (e.g. from photovoltaics, solarthermal power plants, wind or hydropower). This will offer new options and opportunities but also new challenges for the future energy system.

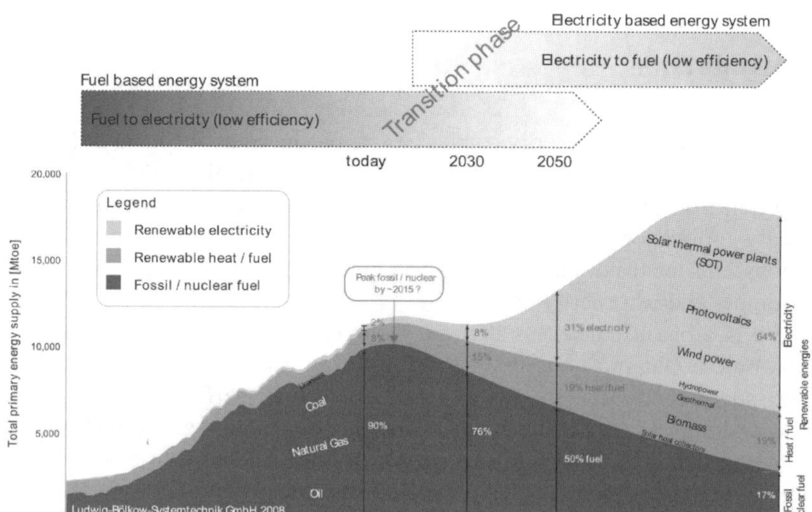

Fig. 1 Future primary energy supply.

To a large extent, the transport sector will be electrified. Soon, road vehicles will start to switch to renewable electricity as a "primary energy source". Hydrogen will become a very essential partner for the future transport sector. As electricity from wind and solar energy is difficult to store, hydrogen will serve as electricity storage and as a transport fuel for future vehicles. Vehicles with longer operating ranges, higher payloads and fast refueling capabilities very likely will have to be hydrogen fuel cell vehicles in the foreseeable future.

3. Technical potential of renewable electricity – worldwide

Renewable electricity could become the most important energy of the world. This figure (Figure 2) summarises the technical potential for the production of renewable electricity from photovoltaics (PV), solar thermal power plants, wind and hydropower by world regions. Today's global electricity use lies in the order of 18,000 TWh per year.

4. Renewable electricity generation

The share of renewable electricity is expected to increase significantly since the largest potentials for renewable energies in Europe and Germany are identified as wind, hydropower, solar and geothermal electricity. In 2007, the German Federal Ministry for the Environment, Nature Conservation and Nuclear Safety has published the Guiding Study 2007 "Development Strategy Renewable Energies"[4]. This study predicts that the electricity production from renewable resources will increase from 74 TWh in 2006 to 156 TWh in 2020. With regard to the current and presumably also the future electricity production this represents more than 25% of the overall production. This scenario also reflects the already enacted and binding goals of the European Union.

5. Renewable electricity storage

With increasing share of renewable electricity our future energy system will require large scale storage systems for electricity. Today, pumped hydro power stations are the only widely used means to store electricity at industrial scale. But the potential for further extension and new installations is very limited. The only technology at present knowledge which has the potential for single storage systems in the 100 GWh range is the storage of hydrogen in underground salt caverns.

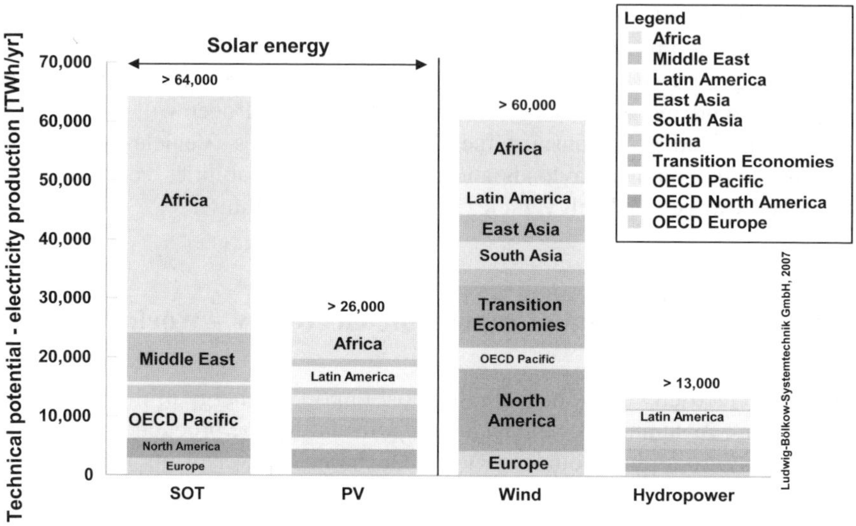

Fig. 2 Potential of renewable electricity.

5.1 Hydrogen as storage for electricity

Hydrogen can be produced from electric power by high pressure electrolysers (≈ 3MPa). For efficient storage hydrogen has to be further compressed before stored in underground salt caverns at a pressure up to 30 MPa (Fig.3). For high power levels the most efficient conversion back to electricity can be achieved in combined cycle power plants. In the lower power range fuels cells can be applied. Roundtrip efficiencies are expected to be in the range of 35- 40 %. The achievable storage capacity of compressed hydrogen in salt caverns is more than one order of magnitude higher than the one of compressed air. The storage of compressed hydrogen in salt caverns being relatively cheap, this technology qualifies especially for long-term storage of bulk energy to be reused during long-lasting unavailability of wind energy.

6. Hydrogen as transport fuel

A further pathway for the utilisation of sustainably produced hydrogen will be its use in the transportation sector. The current discussions on CO_2 reduction and the availability of fossil fuels show, that car industry is speeding up with their efforts to develop hydrogen powered vehicles and that first vehicle sales can be expected for 2010 followed by a broad market entry from 2015 onward [5].

Furthermore, it can be observed that hydrogen propulsion in passenger cars is a twice as efficient end-use technology than today's direct injection engines and thus displaces conventional fuels and powertrains more efficiently than using hydrogen in stationary conversion units (combined cycle power plants or fuel cell systems) where it competes for the time being with almost as efficient natural gas-based end-use technologies.

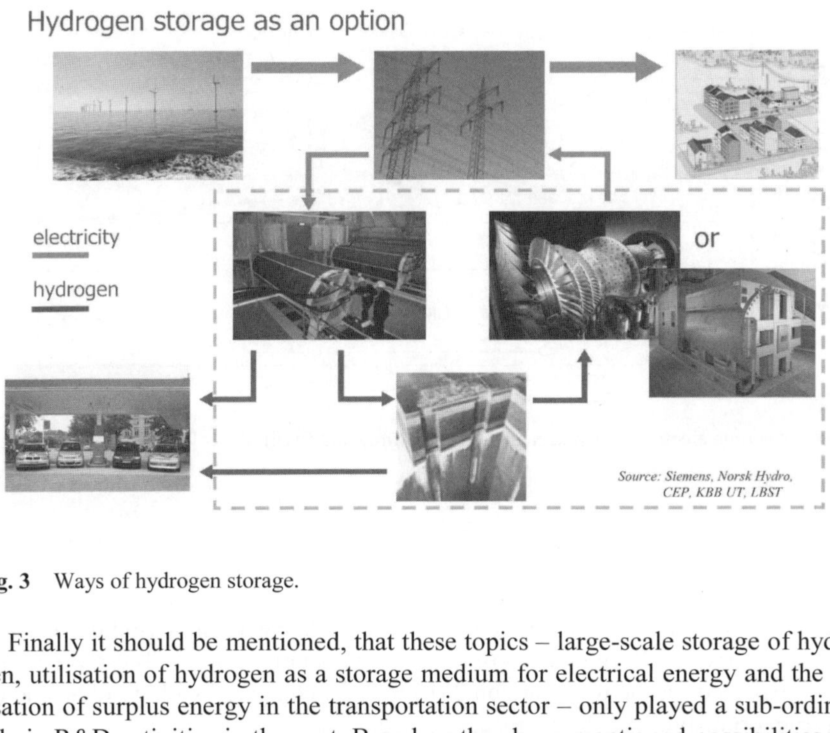

Fig. 3 Ways of hydrogen storage.

Finally it should be mentioned, that these topics – large-scale storage of hydrogen, utilisation of hydrogen as a storage medium for electrical energy and the utilisation of surplus energy in the transportation sector – only played a sub-ordinate role in R&D activities in the past. Based on the above mentioned possibilities this technology can provide, it is urgently needed that both industry and politics attend to this topic and provide the required resources allowing to find suitable answers to the pending changes in the energy landscape.

7. Production costs

Energy costs are expected to increase during the next decades due to the depletion of fossil and nuclear energy sources and rising investments in new power plants and infrastructure.

The graph in Figure 4 shows rising costs for fossil energy sources and decreasing costs for renewable energies depending on the assumption that the break-even point between fossil and renewable electricity production will occur sometime after 2020. Up to this date, the introduction of renewable energies will lead to higher average energy costs whereas after passing the break-even point, the growing contribution of renewable energy sources will reduce electricity costs compared to a purely fossil scenario.

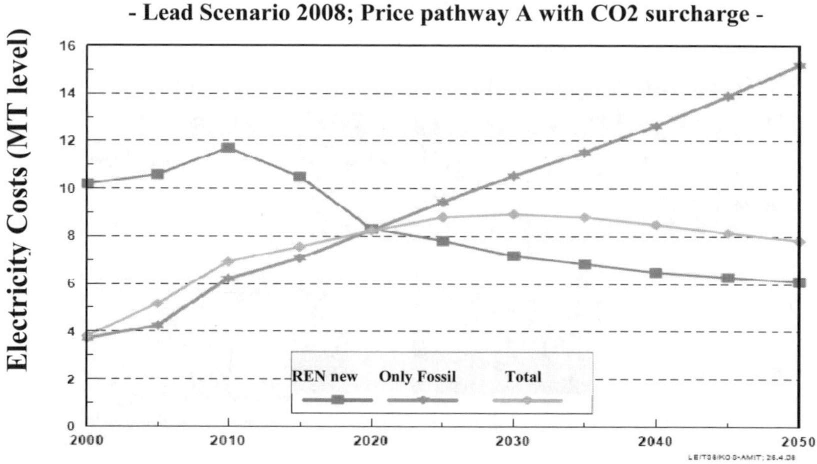

Fig. 4 Average electricity cost scenario for renewable and fossil plants [6].

7.1 Costs of transport fuel

Figure 5 shows the cost of hydrogen from renewable electricity (wind power, photovoltaics) and BTL (biomass-to-liquid) from short rotation forestry (farmed wood) expected for the time horizon 2010 to 2020 compared with gasoline and diesel from crude oil.

The fuel costs for BTL are about the same as for hydrogen from renewable electricity if the higher efficiency of fuel cell vehicles is taken into account. Both BTL and hydrogen from renewable electricity lead to fuel costs which are slightly below those of taxed conventional gasoline and diesel at a crude oil price of 50 €/bbl. In the future, increasing prices for crude oil will lead to an increase of the price for conventional crude oil based fuels. Cost parity for hydrogen with untaxed conventional gasoline and diesel will result.

Renewable transportation fuels such as BTL and hydrogen from renewable electricity can reduce GHG emissions significantly. On the other hand, potentials for biomass are limited and thus also the potential for BTL.

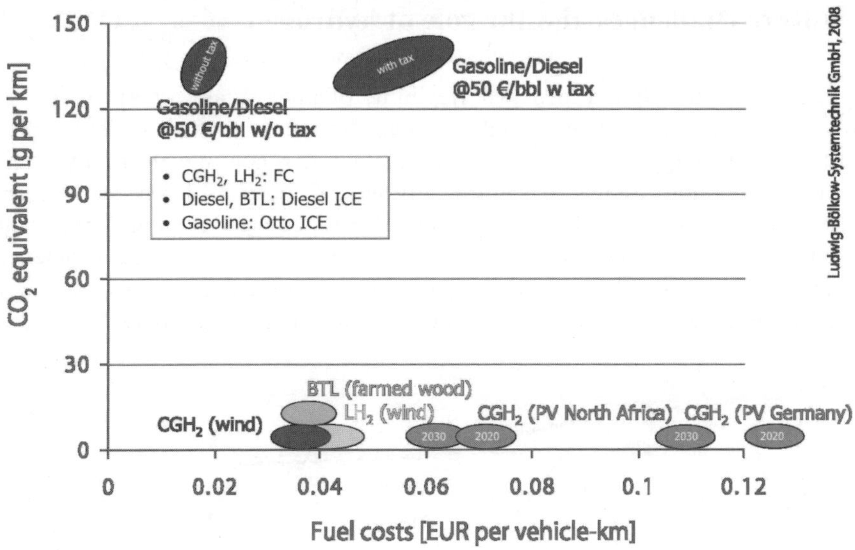

Fig. 5 Well-to-wheel fuel cost versus CHG emissions[1].

8. *Efficiency chains of different technologies*

Fig. 6 indicates the efficiency chains of different fuel and powertrain technologies.

The best overall efficiency is reached by an electric vehicle with a battery storage. But it should be mentioned, that in this case the range of a car is limited to about 100-200 km with a refilling time of several hours, shorter filling times only for partial battery loads and consequently shorter ranges.

A hydrogen car with fuel cell powertrain and CGH_2-storage (70MPa) reaches a well-to-wheel efficiency of about 28%, representing the second best value of all investigated chains.

---

[1] Major assumptions of Figure 5: The fuel consumption of the hydrogen fuelled fuel cell hybrid vehicle is assumed to be 0.84 MJ/km (~ 2.6 l gasoline equivalent per 100 km). The fuel consumption of the gasoline fuelled hybrid vehicle amounts to about 1.62 MJ/km (~ 5.0 l gasoline equivalent per 100 km) and the fuel consumption of the diesel fuelled hybrid vehicle amounts to about 1.46 MJ/km (~4.5 l gasoline equivalent per 100 km or ~4.0 l diesel per 100 km).The fuel consumption for the vehicle is derived from Concave/EUCAR/JRC [7]

9. Future challenges and the role of hydrogen

Hydrogen with the highest feedstock flexibility of any available energy carrier can provide a clean storage media and facilitate the transition from today's energy world into an energy world with growing fluctuating renewable electricity supply. Hydrogen therefore allows to significantly extend the use of renewable energy sources in the transport sector. In particular, hydrogen used as an automotive fuel in hybridised fuel cell powertrains can provide superior well-to-wheel efficiencies compared to today's fuels and thus reduce primary energy consumption and greenhouse gas emissions while maintaining today's operating ranges.

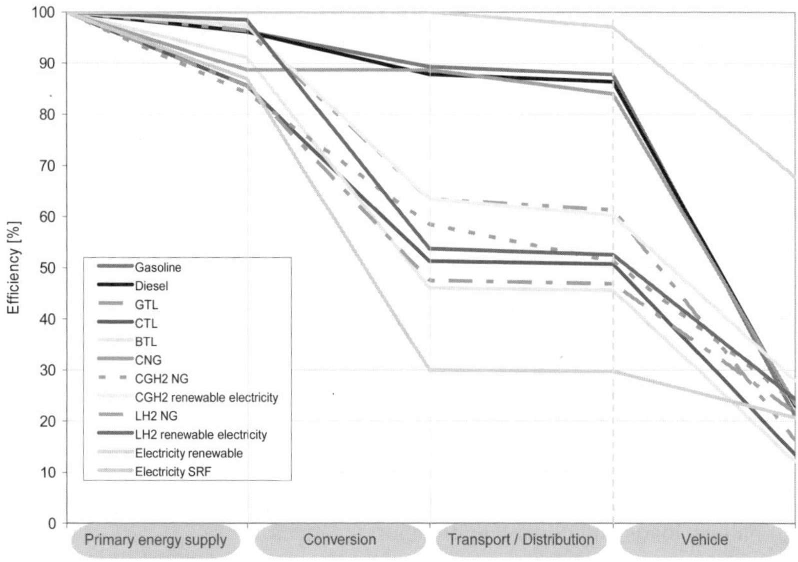

Fig. 6 Efficiency chains of different fuel/vehicle technologies.

9. Future challenges and the role of hydrogen

Hydrogen with the highest feedstock flexibility of any available energy carrier can provide a clean storage media and facilitate the transition from today's energy world into an energy world with growing fluctuating renewable electricity supply. Hydrogen therefore allows to significantly extend the use of renewable energy sources in the transport sector. In particular, hydrogen used as an automotive fuel in hybridised fuel cell powertrains can provide superior well-to-wheel efficiencies compared to today's fuels and thus reduce primary energy consumption and greenhouse gas emissions while maintaining today's operating ranges.

References

1 Energy Watch Group, Crude Oil – The Supply Outlook, Oct. 2007, EWG-Series No 3/2007, http://www.energywatchgroup.org.
2 Energy Watch Group, Coal: Resources and Future Prod., March 2007, EWG-Series No 1/2007.
3 Energy Watch Group, Uranium Resources and Nuclear Energy, Dec. 2006, EWG-Series No 1/2006.
4 J. Nitsch, "Lead Study 2007", DLR Institute for Technical Thermodynamics, February 2007
5 Joint Press Release of Linde, Daimler, EnBW, NOW, OMV, Shell, Total and Vattenfall: Initiative "H2 Mobility" - Major companies sign up to hydrogen infrastructure built-up plan in Germany, 10 Sept. 2009.
6 J. Nitsch, "Lead Study 2008", DLR Institute for Technical Thermodynamics, October 2008
7 CONCAWE, European Council for Automotive R&D (EUCAR, Joint Research Center (JRC); Well-to-Wheel Report, March 2007; http://ies.jrc.ec.europa.eu/WTW.

EFFICIENCY IMPROVEMENT OF INTERNAL COMBUSTION ENGINES BY WASTE HEAT RECOVERY WITH RANKINE CYCLE AND AN ADVANCED TURBOCHARGING PRINCIPLE

A. Gotter[1], E. Küpfer[2]

[1]gofficient, An der Schmackertz Kull 9, D-41564 Kaarst, Germany
Email: info@gofficient.de

[2]Innodrive GmbH, Karl-Brand-Straße 68, D-9742 Schweinfurt, Germany
Email: ewald.kuepfer@innodrive.net

Abstract Internal Combustion Engines have a maximum efficiency of approx. 40% and, if operating at part load, much lower. Nearly the same amount of produced mechanical energy is wasted by the exhaust gases and a further major amount of energy is wasted by cooling the engine. By that fact, using the Waste Heat with a Waste Heat Recovery process seems to be a great option to increase overall efficiency of the engine. This also helps saving fuels and emitting less carbon dioxide. As there are several systems available, which can convert the thermal energy in the exhaust gases, this article gives an overview of the available technologies and gives an evaluation for different applications, power generation and mobility. The principles of the most promising technologies are described and explained. This is on the one hand an improved turbocharger technology and on the other hand, a steam based Rankine cycle, which combines high efficiency with moderate costs.

1. History of Engine Technology Development

Internal combustion engines are used in millions in passenger cars, commercial vehicles and also in stationary block heat and power plants. In the most usual applications – passenger cars – the actual power demand is much lower than the maximum nominal power of the engine.

A car's fuel consumption is determined according to the NEDC (new European driving cycle) standard in various tests. For most vehicles, the complete test cycle is operated in part load. In the diagram (figure 1), these operating points are shown in the fuel consumption map of a typical Otto engine as well as additional operating points relevant for a sportive driver.

J. Wellnitz et al. (eds.), *Sustainable Automotive Technologies 2010*,
DOI 10.1007/978-3-642-10798-6_18, © Springer-Verlag Berlin Heidelberg 2010

As can be seen, most operating points are in an area of high specific fuel consumption with BSFC (brake specific fuel consumption) >300 g/kWh, while the point of the best fuel consumption never was reached.

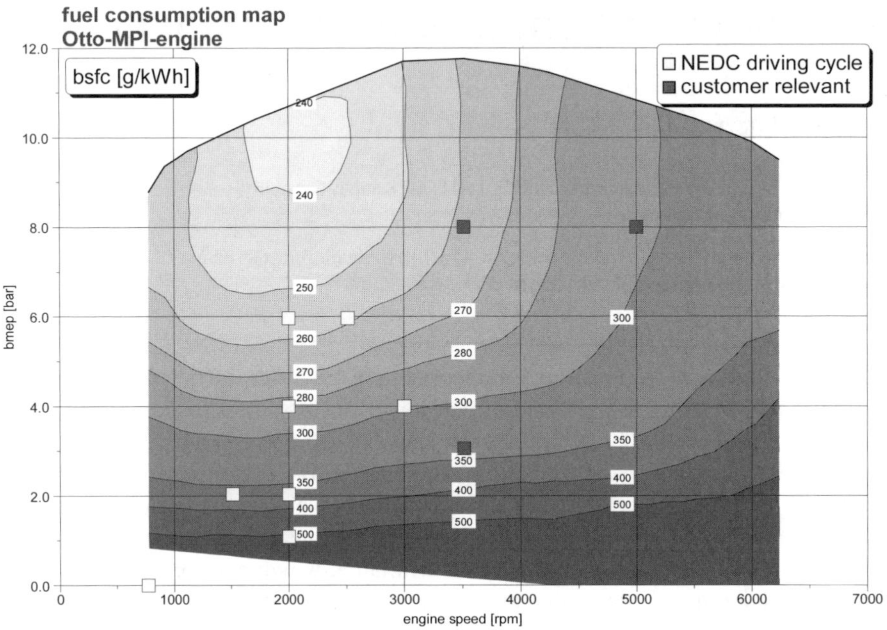

Fig. 1 Typical bsfc map of a gasoline engine.

This is the reason for the development of part load improving technologies, such as variable valve trains and stratified direct injection. Also friction reduction has a major impact on the fuel economy in the lower part load. This development of modern passenger car engines reduced the fuel consumption by up to 30% in the last 20 years. As result of all these technical innovations, the part load efficiency did increase, as shown in the next diagram (figure 2):

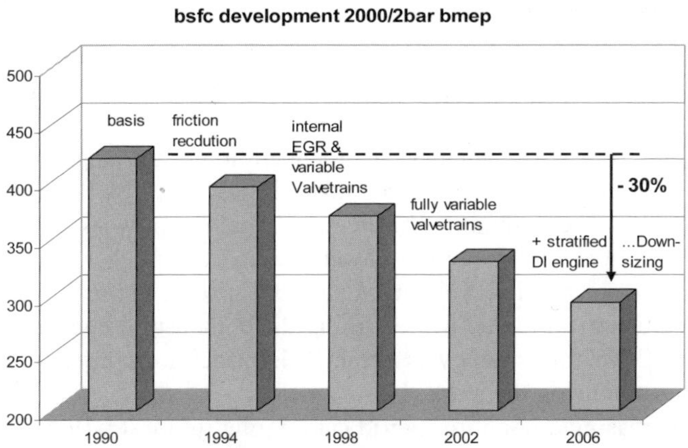

Fig. 2 Historical BSFC part load development of modern vehicle gasoline engines.

However, the fuel consumption in the optimal operation point was not affected by these technology development. The reason can be found in the principles of e.g. DI injection or variable valve trains. These concepts avoid load change pumping losses in the part load, while at higher load, there is nearly no pumping loss anyway. A slight benefit did impact the higher loads, caused by increased compression ratios of direct injection engines and friction reduction, which also has a small efficiency effect at higher load. By that, the fuel consumption at the high loads decreased by ~ 5%.

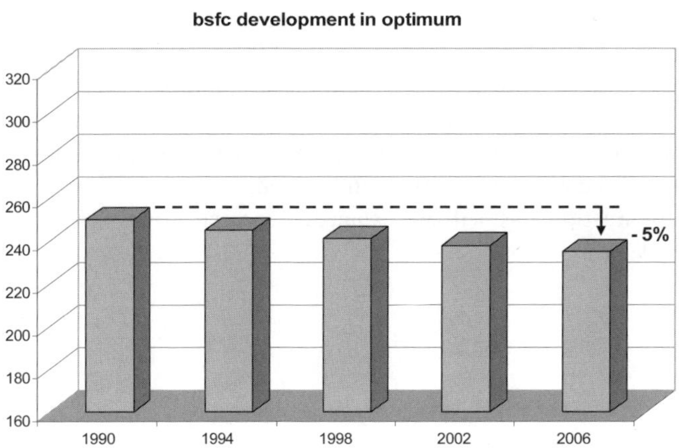

Fig. 3 Historical BSFC optimum development of modern vehicle gasoline engines.

1.1 Future automotive trends

There is one clear trend in the development of automotive drive technologies: Hybridization.

Of course, one Hybridization benefit is the recuperation of braking energy. But the second, even more effective benefit of hybrid drives is the movement of the operating point closer to the optimum. The electrical engine can work as generator at some driving conditions and move the power demand to a higher engine load, while driving with low or medium velocity. Later, the saved electrical energy can be used for driving electrical alone, while turning the combustion engine off. The fuel economy benefit of this strategy is great. This principle is already used in many hybrid cars available on the market.

Assuming an increasing amount of hybrid cars, in the future it will be less important to optimise part load efficiency, but it becomes really urgent to improve optimum efficiency of internal combustion engines. Commercial vehicles and stationary engine operate already today much closer to the optimal operating point.

2. TwinAV-principle

The pressure at the end of the expansion phase has still some 5–8 bars. While opening the exhaust valve against ambient pressure, this pressure energy is wasted.

If the exhaust pressure will be increased and the gas will expand through a turbine, such as realised with a turbocharger system, a part of that energy is used to drive the turbocharger. However, also a static backpressure will occur, which leads to pumping losses while the exchange phase of the cylinder.

The idea of the TwinAV-concept is to avoid these pumping losses by different actuation of both exhaust valves: The first exhaust valve opens at the end of the expansion phase and conducts the gas to the turbine with static backpressure, so that a great amount of the pressure energy can be used.

The residual combustion gas will be exhausted via the other exhaust valve – against ambient pressure.

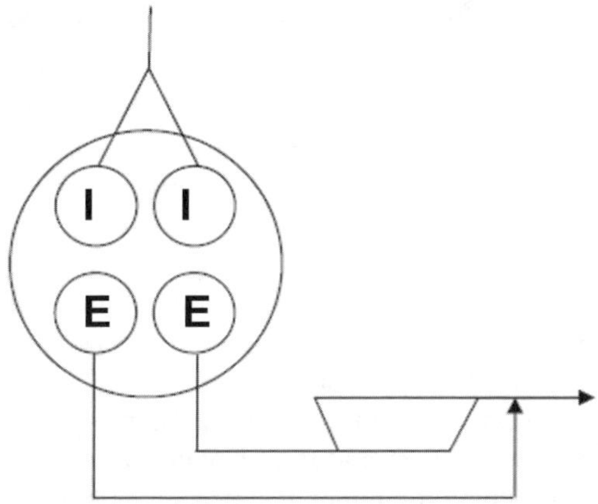

Fig. 4 Principle scheme of TwinAV hardware configuration with Intake (I) and Exhaust (E) valves.

Numerical simulations have shown, that the mass fraction going to the turbine in this application is approx. 50–70% of the total mass fraction, depending on turbine backpressure and cylinder pressure.

In contrast to conventional turbocharger set up, there is no penalty increasing the backpressure by a smaller turbine. By that fact, the turbine can be sized smaller, so that the won energy at the turbocharger is higher than the one with conventional setup, even at only 50% mass flow fraction through the turbine.

As a result, at high engine loads, an efficiency increase of 3 to 7 % can be achieved by eliminating the static backpressure. Even at part load at medium and higher engine speeds, an efficiency increase can be realised. While a conventional turbochargers causes a backpressure also in those part load operating points, there is no backpressure with the Twin-AV principle.

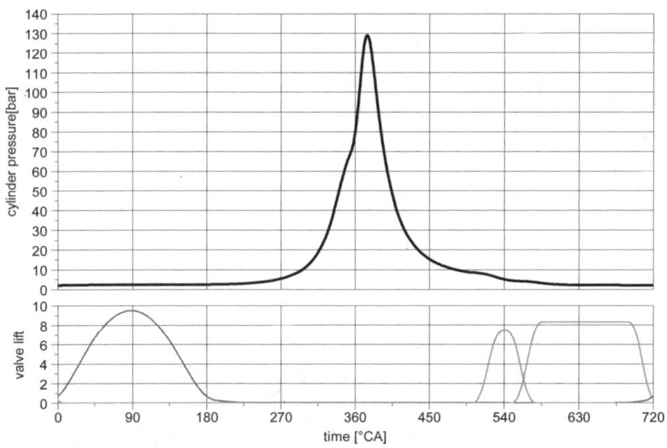

Fig. 5 Typical pressure and valve lift over degree CA for TwinAV principle.

However, there are also two limiting factors for the TwinAV concept:

1. Due to the higher pressure upstream turbine, the exhaust gas temperature at the turbine inlet will be up to 80–100°C higher than the one of a conventional turbocharger system. By that fact, it is strongly recommended to use 1050°C turbocharger technology or a lean combustion principle, such as natural gas or Diesel.
2. Due to the short event time for the exhaust valves (~110–160°CA), the application for high-speed engines concepts is limited by high valve actuating forces at high engine speeds. Engines speeds of 5000rpm can be realised, higher speeds need special attention to the valve springs.
3. Concluding, the TwinAV concept is the ideal charging concept for stationary applications, such as block heat and power plants and for future passenger car drives, where the combustion engines is designed as hybrid drive. The fuel efficiency benefit is up to 7%.

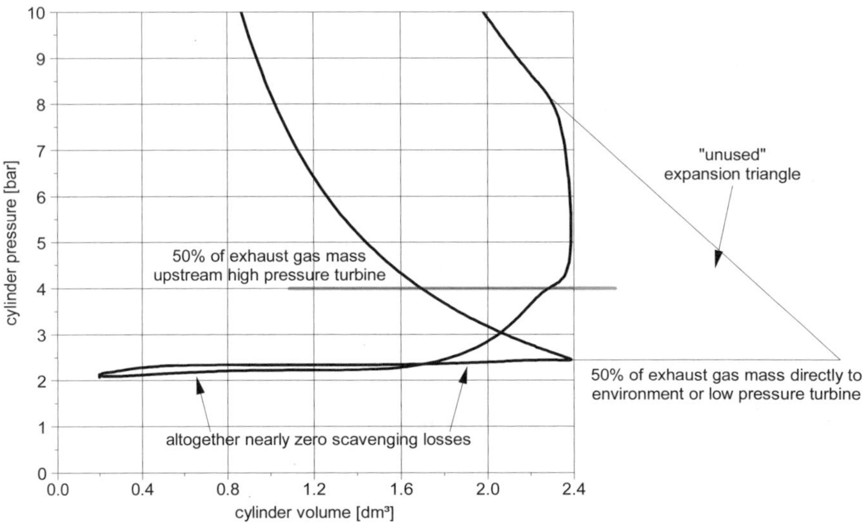

Fig. 6 pV-diagram of load exchange for TwinAV principle.

3. Waste heat recovery system

The major part of the fuel energy is wasted in the exhaust gases and in the cooling water.

If balanced at the exhaust valve, approx. 35-40% of the total fuel energy is thermal energy of the exhaust gas, while 20% is wasted in the cooling water. This ratio is moved a bit by cooled exhaust manifold and turbocharger, so that the ratio cooling water energy / exhaust gas energy is nearly 1:1 in most applications.

However, the exhaust gas has a temperature of more than 400°C, which could be converted with an efficiency of more than 50%, if Carnot's law is applied. However, that value can never be reached. But there are many technologies available to convert the energy of a hot exhaust gas stream, as summarized in the figure below.

One special aspect has to be considered, if exhaust gas should be used for the energy conversion: The temperature of the gas is not constant (as the output from a concentrating solar receiver, for example). That means, the working fluid must be heated starting from a low temperature with ideally linear characteristic. If that can be solved perfectly, the exhaust gas temperature at the heat exchanger outlet is very low, otherwise only a small amount of the total exhaust energy can be used for heat recovery. By that reason, a Rankine cycle is the best solution for exhaust gas recovery.

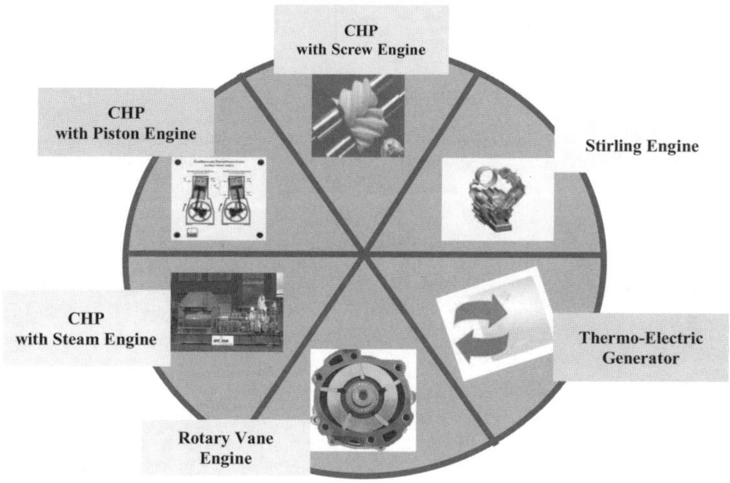

Fig. 7 Available technologies for waste heat recovery systems.

3.1 Working fluids

There are several working fluids available: The most common working fluid for a Rankine cycle is water, which is used in large power plants. Also very common are organic working fluids, such as Isopentane, used in a ORC cycle (organice rankine cycle). Also feasible working fluids are alcohols and ammonia.

One limiting factor for the effective usage of a Rankine cycle is the maximum pressure. This is the main parameter for the design of the turbine and the heat exchanger. A pressure limitation of ~25bar seems to be strongly necessary, whereas increased system pressure boosts the investment disproportionately.

For a selection of different working fluids the next figure gives an estimation of cycle and overall efficiency. Basis for all processes is a common exhaust gas temperature of 400°C, with ~20°C temperature drop within the heat exchanger.

working fluid	pressure [bar]	process efficiency	total efficiency
water	150	31,00%	19,53%
water	25	24,30%	15,31%
Ethanol	110	25,70%	25,70%
Ethanol	25	19,70%	18,32%
Methanol	100	24,70%	24,45%
Methanol	25	18,00%	16,56%
NH3	320	23,80%	23,80%
Isobuhtane	200	14,60%	14,60%
Isopenthane	160	18,00%	18,00%

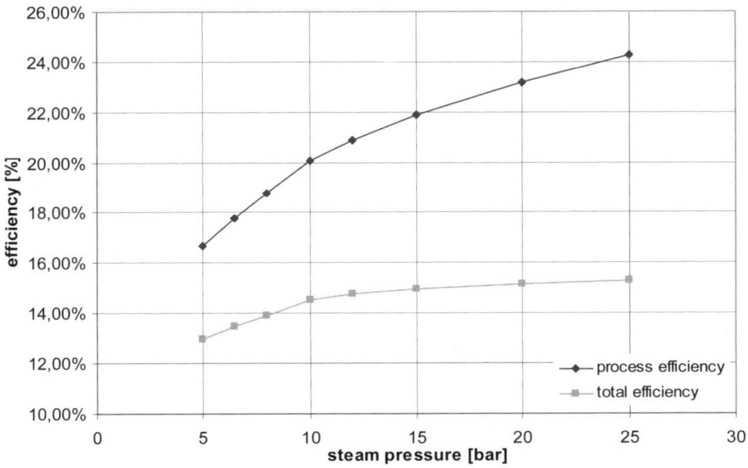

Fig. 8 Property comparison for different waste heat recovery fluids and efficiencies for water depending on steam pressure.

Here, water is the best working fluid, closely followed by some alcohols. However, water is not hazardous and there is most experience with water as working fluid, so the decision for water is set.

3.2 Proposed system design

Two disadvantages of water are the high necessary pressure ratio for expansion and water droplets while expanding into the wet steam area. For most turbine, this would be life-limiting.

Based on that fact, we propose a two-stage screw expander system. The system schematic will look like shown on the next figure.

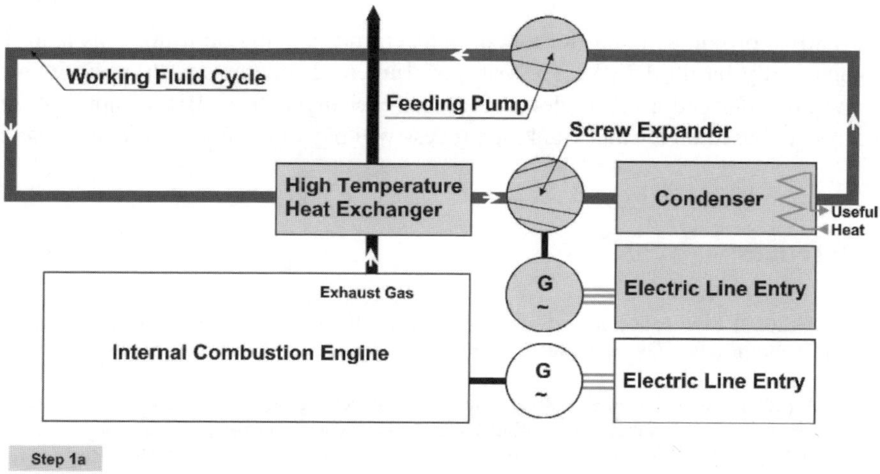

Fig. 9 Proposed first step configuration for our prototype power plant.

An important component especially for mobile use is the exhaust heat exchanger. It should be very small with a high heat transfer rate per square meter. There we found a manufacturing method and a functional surface in order to force the steam building. Both, the functional surface and the new manufacturing method promise very good results as shown below.

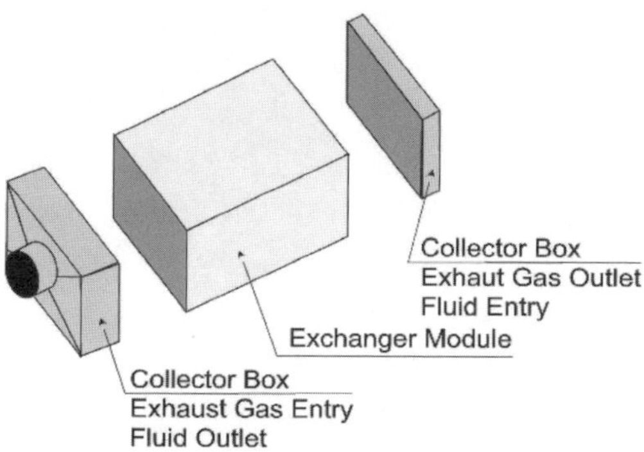

Fig. 10 Exhaust gas heat exchanger.

The great benefit of this application is the well-known und tested parts of the whole system. By that, the technical risk of the system is very low, while the efficiency of the exhaust gas cycle is up to 18%. So a efficiency gain of 11%, relatively to the main engine can be achieved.

3.3 Prototype and reference location

Currently a prototype plant, located at Schweinfurt is being built-up. This plant is funded partly by the BMWi, German government. It operates with an optimised screw expander and a newly developed heat exchanger. With Biogas application, the system based on a water Rankine process will pay out within less than 2 years.

References

Gotter A (2009) Effiziente Energiewandlungsmaschine durch kombinierte Brennkraftmaschine mit Dampfprozess. German Patent Application DE 102007038073A1, published 19 Feb 2009.

Gotter A (2006) Potenzial einer vollvariablen Ventilsteuerung zur Verbesserung des Betriebsverhaltens eines Ottomotors mit Aufladung. RWTH Aachen, Germany.

Ringler J, Seifert M, Guyotot V, Hübner W (2009) Rankine Cycle for Waste Heat Recovery of IC Engines, *SAE Paper*, 2009-01-0174.

Habermann K, Lang O, Salber W, Wohlberg R, Gotter A (2005) Ventiltriebsvariabilitäten für aufgeladene Ottomotoren. *10. Aufladetechnische Konferenz*. Dresden, 22-23 September 2005. Dresden, Germany.

Part 5
New structures and materials

CELLULAR TITANIUM A BIONIC MATERIAL FOR AUTOMOTIVE AND AVIONIC APPLICATIONS

M. Hartmann[1] and J. Wellnitz[2]

[1]Private Institute of Technology and Artistic Design (ITD E.V.)
Marie-Curie-Straße 6, 85055 Ingolstadt, Germany
E-mail: joerg.wellnitz@fh-ingolstadt.de

[2]University of Applied Sciences,
Esplanade 10, 85049 Ingolstadt, Germany
E-mail: michael.hartmann@itd-in.de

Abstract The challenge of developing new sustainable mobility concepts cannot only be reduced to the development of alternative powertrains. The weight reduction is also a very essential part of this. In view of this, a cellular titanium material for different applications in the automobile, avionic or medical industry is going to be investigated in the "Anthrocells-Project". To give a short overview about the mechanical properties of this open-pore titanium foam, the results of the quasi-static compression and tension test are summarized in this paper. Furthermore the influence of the porosity to the stiffness and strength is shown.

1. Introduction

Despite of all efforts and activities in the field of cellular materials over the last decade, the number of concrete applications in automobile sector, aviation or architecture is low till today.

After Lefevre (2008) the most important commercial applications of porous metals and metallic foams remain open-cells materials for the production of filters, gas-flow con-trolling devices, batteries, catalytic converters, bearings and for a short time in the biomedical industry. These applications are characteristically for the current problem. The lack of *reproducible* material data and the limited manufacturing processes prevent the efficient usage of cellular materials for structural applications. Which means, that the economically production of these materials with the required quality and reproducibility is not sufficient.

To counteract this problem the "Anthrocells-Project" is going to research the possibilities of the production of a material with strength and stiffness optimized structures, like the nature. First step, to realize this aim was the production of an open-pore titanium foam, using the space holder method and tested under quasi-

J. Wellnitz et al. (eds.), *Sustainable Automotive Technologies 2010*,
DOI 10.1007/978-3-642-10798-6_19, © Springer-Verlag Berlin Heidelberg 2010

static loads. Furthermore samples with different porosities were manufactured to investigate the influence of the porosity to the stiffness and strength of the material.

2. Manufacturing process

After Rausch is the melt-metallurgical manufacturing of porous titanium because of the high melting point (1600 °C - pure titanium) and the high reactivity of molten titanium interlinked with high efforts in the process control.

Powder-metallurgical techniques have the advantage of low process temperatures and a great variety of design possibilities. In the last years a number of powder-metallurgical manufacturing processes have been developed with different success. Two of the most interesting ones are the space-holder method, described by Bram, and the laserforming technique after Wirths. Because of the availability and potential we have chosen the space holder technique for our investigations.

The specimens, shown in Figure 1, were produced by the IFAM-Bremen (Fraunhofer Institute for Manufacturing and Advanced Materials).

Fig. 1 Specimen for tensile and compression tests produced with the space holder method.

The production involves the following steps:

- Mixing of powder and space holder materials
- Pressing of a green body
- Removal of the space holder
- Sintering

In the first step the titanium powder, a TiAl6V4-ELI "extra low interstitial" alloy and the paraformaldehyde - space holder are mixed. The grain size of the space holder particles is smaller than 500 μm and the grain size of the titanium powder is 22 - 45 μm respectively 4 - 45 μm.

In the next step the mixture is compacted and the space holder is removed. The ejection of the paraformaldehyde is a catalytic process under nitrogen atmosphere at 110°C. At last, the parts are sintered in an oven under argon atmosphere at 1300°C for 1 hour.

3. Experimental work

To investigate the mechanical properties, static compression and tension tests were carried out on the University of Applied Sciences Ingolstadt.

3.1 Tension

For the tension tests, the Zwick-Z010 universal testing machine with 10 kN maximum force was used. The specimens, shown in figure 1, were manufactured according DIN EN ISO 2740. The cross sections of the bone-similar samples were 5 x 6 mm. The parallel length was adjusted at the machine to 25 mm and the testing-speed was set to 0.025 mm/s.

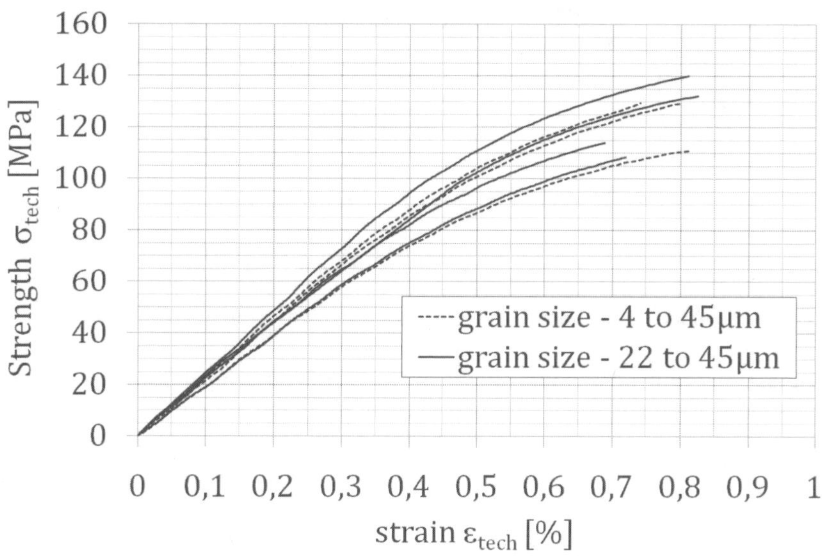

Fig. 2 Test results of the tension test on bone-similar specimen with different grain sizes. The porosity of the samples is 50 Vol.%.

The stress-strain curves, depicted in Figure 2, show a relatively homogeneous behaviour. The average values for the Young's Modulus and the strength, of both specimen groups is nearly the same. The Young's Modulus is $E = 21$ GPa and the average tension strength is $\sigma_{0,2} = 122$ MPa.

3.1 Compression

The compression tests were realized on a ZWICK-Z250 testing machine. The cylindrical specimen, with different porosities (50 Vol.%, 55 Vol.%, 65 Vol.%) were compressed between two parallel plates. The maximum force of this machine is 250 kN. The diameter of the samples was 17.5 mm and the height was 25 mm.

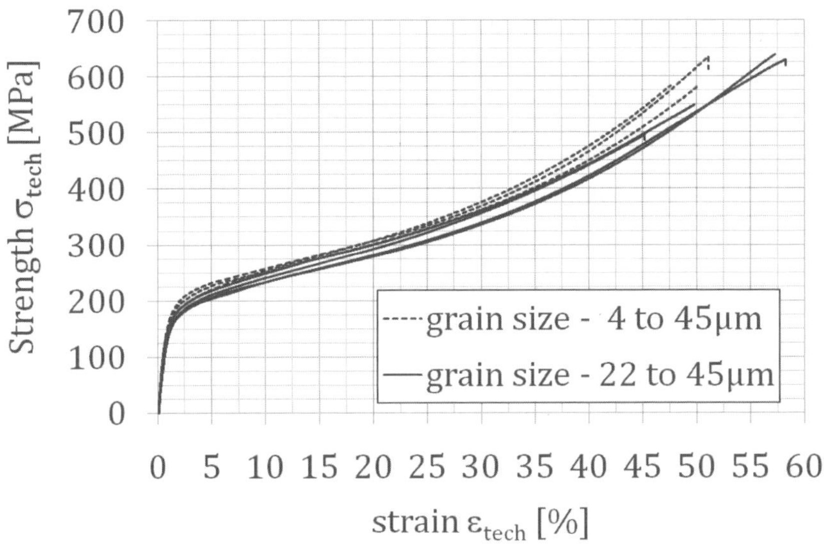

Fig. 3 Results of the compression tests on samples with porosity of 50 Vol.%.

The stress-strain behaviour, shown in Figure 3, is typical for cellular metals under compressive load. The influence of the powder material can be detected through the comparison of the mechanical properties. The Samples with the 4 – 45 μm particle size shows a stiffer behaviour than the specimen with the granulate size 22 - 45 μm. In the strength value, the results offer inverse values. The reason for this is the pore to pore connection and is explained more accurately by Hartmann.

Porosity

To evaluate the influence of the porosity to the stiffness and strength, specimens with higher porosities are tested with the same testing parameters. The summary of the results is shown in the following Figure 4.

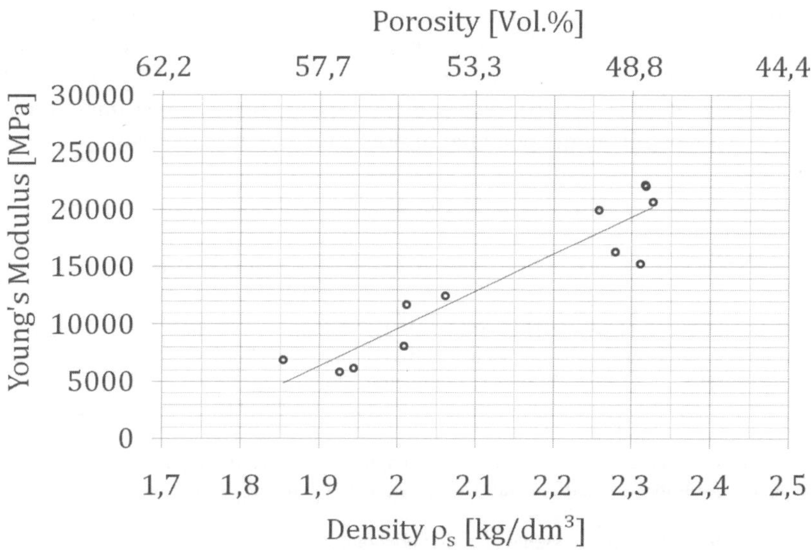

Fig. 4 Correlation between the Young's Modulus and the Density for compression tests. In the diagram are only the specimens with the particle size 4 - 45μm considered.

The compression tests revealed a linear correlation between the Density and the Young's-Modulus respectively a dependency between Density and strength. This relation is also mentioned in Gibson et al (1997).

4. Conclusion

The accomplished experimental investigations are carried out on powder-metallurgical manufactured open-cell titanium foam with porosities of 50 Vol.%, 55 Vol.% and 65 Vol.%. They serve as a basis for the development of a new lightweight structure optimized porous metal for various applications in the automobile and aviation industry.

The tests have shown that the mechanical properties under quasi-static load could be affected by the porosity. By choosing the adequate density the desired mechanical properties can be aligned over a wide range. The stiffness of the investigated foams can be adjusted from 5 – 25 GPa. The strength varied between 50

MPa and 150 MPa. Furthermore it could be shown, that the material is optimal used in compression.

The variation of the grain size shows only a small influence on the stress-strain curve of the material.

With respect to fatigue behaviour no representative results are available, till now.

Besides the applications in the aviation and automobile sector there are also possibilities in the medical industry. A comparison between the detected values of the foam and the mechanical properties of bones, mentioned by Reilly et al. (1975) and Hayes et al. (1976), revealed similar properties. Some implant manufacturers have recognized the potential of the cellular material already.

References

Bram M., Laptev A., Buchkremer H.P., Stöver D.: Herstellung von hochporösen, entkonturnahen Titan-Formkörpern für biomedizinischen Anwend-ungen, In: Mat.-Wiss u. Werkstofftechnik, WILEY-VCH Verlag, 35/4, 213 -218, Weinheim, 2004

Gibson L. J., Ashby M. F.: Cellular solids _ Structure and properties. 2nd edition.Cambridge University Press, Cambridge, 1997

Hartmann M.: Ein Beitrag zur Lebensdauerbetrachtung von Schaumwerkstoffen für Anwendungen in der Automobil- und Luftfahrtindustrie, Universität des Saarlandes, Dissertation, in press.

Hayes W., Carter D.: Postyield Behaviour of Subchondral Trabecular Bone, In: Journal of Biomedical Material Research 7, 1976

Lefevre L.P, Banhart J., Dundand D.C.:Porous Metals and Metallic Foams: Current Status and Recent Developments, In: Advanced Engineering Materials, Wiley-VCH Verlag, 10/9, 775-787, Weinheim, 2008

Rausch G., Hartwig T., Weber M., Schulz O.: Herstellung und Eigenschaften von Titanschäumen, In: Mat.-Wiss u. Werkstofftechnik, WILEY-VCH Verlag, 31/2000, 412 -414, Weinheim, 2000

Reilly D., Burnstein A.: The elastic and ultimate properties of compact bonetissue, In: Journal of Biomechanics 8, 1975

Wirths T.P.: Herstellung von Knochenimplantaten aus Titanwerkstoffen durch Laserformen, Rheinisch-Westfälische Hochschule Aachen, Dissertation, 2005

RESEARCH AND DEVELOPMENT OF A NEW AND SUSTAINABLE COMPOSITE: "NATURAL STONE LAMINATE"

L. Müller[1], J. Wellnitz[1,2]

[1]Private Institute for technique and artistic design e.V.,
Marie Curie Straße 6, 85055 Ingolstadt,Germany,
www.itd-in.de, lars.mueller@itd-in.de

[2] University of applied sciences Ingolstadt,
Esplanade 10, 85049 Ingolstadt, Germany,
www.haw-ingolstadt.de. joerg.wellnitz@haw-ingolstadt.de

Abstract NSL (Natural Stone Laminate) is a new sandwich material combining the favorable properties of natural stone and metal. Using two widely available and low priced components, the laminate will reach a competitive price level. The fields of application include construction, security and design. The current research is mainly working on the optimisation by running several tests. Finding different values for Tensile strength, Young's Modulus Bending- and impact behaviour are just a short overview of our work. Further development especially in the fields of forming and processing NSL will certainly open up even more possibilities for this competitive sandwich material. By regulating the used layers and materials NSL perfectly fits to the consumers needs.

1. Introduction

Light construction is the philosophy of gaining maximum reduction in weight. Especially in times of raising ecological damage, R&D efforts are on their way to find *lighter constructions*. On the one hand this is achieved through weight optimization of existing materials, on the other hand through engineering *totally new materials*. Variable sandwich- and laminate-structures combine different materials in order to *create optimized material behavior*. Besides material properties, *availability and price* play a major role in the development process of new materials.

Strengthening plastics using carbon fibre herein surely was one of the most interesting ideas in former times. The developed combinations turned out to be suitable for usage and are certainly very useful in their field of application. Also combinations of plastics and glass fibre are well known, and GLARE® was a further development in using aluminum. Figure 1 shows the material mix in a modern airplane and underlines the extensive usage of composite structures.

J. Wellnitz et al. (eds.), *Sustainable Automotive Technologies 2010*,
DOI 10.1007/978-3-642-10798-6_20,

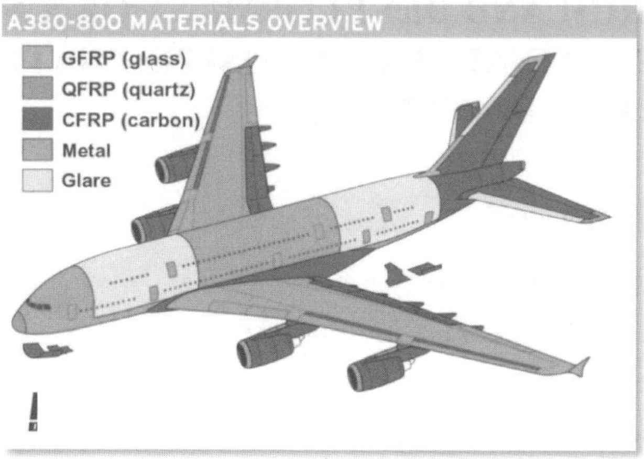

Fig. 1 Material overview in long distance airplanes [FlightGlobal, 2009].

2. Approach

To consider availability and price being the major focus in developing *a new material* a new idea needs to be found.

In the field of metallic materials a lot of realisable variants have been created and are still in use. Especially in *modern Airplanes* aluminium is often used as a light and stable metal in a lot of fields. The problem of difficulties in *availability* is negligible and one can count on the experiences in using aluminium and steel as it creates reliable and save constructions. To minimize damage in cases of debris containment a combination with an impact resistant material is very rewarding and surely opens interesting fields of application.
Earth is a terrestrial planet, meaning that it is a rocky body. It is the largest of the four solar terrestrial planets, both in terms of size and total mass. Considering availability issues stone as a new *"material"* is quiet obvious.

The "*Private Institute for technique and artistic design e.V*" and the "*University of Applied Sciences Ingolstadt*" are intensively working on developing a *brand new composite*. A thin foil of stone is brought to a metal carrier which is the source for creating a laminate of several coatings. The result is called "*Natural Stone Laminate*" (NSL – Figure 2) and *combines advantages* of both of the different elements.

Fig. 2 Natural stone laminate.

By changing the properties of the three constituents - metal, stone and the way of bonding - the behaviour of the laminate can be adjusted to the users' requirements. There is no need for temperature treatment on the material and no difficulties of machining. By using high temperature resistant adhesives it even could be used in hot environments *(up to 500°C and higher).*

A density of 2,7 kg/dm³ with simultaneously very good *bending behaviour, acoustic damping, impact resistance and many more*, the NSL reaches different fields of the market.

3. Production

The pilot production is laboratory work. Layers of an "AB" combination - consisting of metal and stone - are produced in different types depending on their needs. By bringing several layers together symmetrical and asymmetrical plates can be created and formed, always allowing for construction issues.

Fig. 3 Schematic of natural stone laminate.

Figure 3 shows the first "ABA" Layer of a *symmetrical assembly*. The used material is a combination of *Bianco Sardo* and *aluminium* using an *adhesive*. Investigation is also working on alternative possibilities of bonding (e.g Ceramic, Ultrasonic) and optimisation of serial production.

4. Experimental and numerical investigations and validation

4.1 Bending behavior

For a better understanding of the materials behaviour several experimental and numerical testing are planned within the projected research. One of the steps is the comparison of a mathematical model with *laboratory work* which is done in cooperation with the *University of Applied Sciences in Ingolstadt*. Within a *three-point-bending test* a specimen is bent while data of force and elongation are collected.

As it can be seen in Figure 4, a compressive loaded *AB – Laminate* reaches distinctly better results considering bending behaviour – stiffness and strength. That's not exciting, knowing the good values of comprehensive strength of the natural stone.

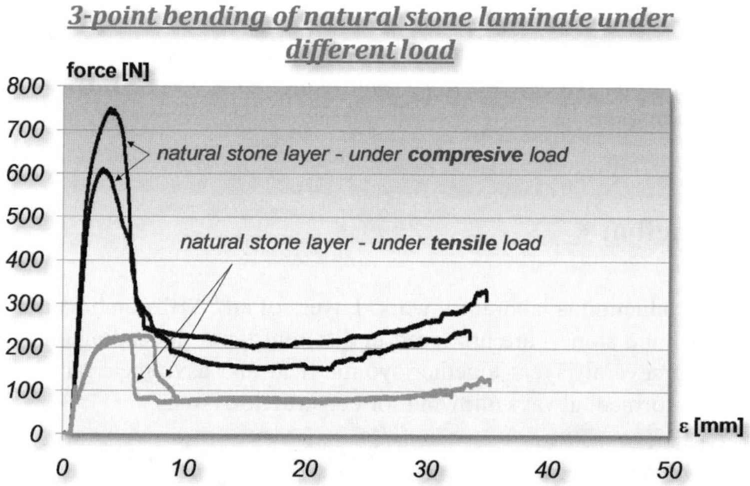

Fig. 4 Bending behavior under selected loads.

Although stone is very brittle, the fact that it is used in thin foils and set between aluminium provides it with a *very good bending behaviour.* That even allows a certain *amount of bending the stone*, without loosing the function of the composite. In Figure 5 that effect is shown.

Fig. 5 layer of NSL using "Bianco Sardo" after bending.

By calculating a *Young's Modulus* of the entire laminate a model can be built for FE methods to integrate the CMA in engineering construction methods. For supporting the theory several laminate theories have been applied to composite laminates. The classical laminate theory is an extension of the Love – Kirchhoff assumption for isotopic plates and can be applied if the laminate is thin because transverse shear stresses are not considered [Altenbach, 1996]. As the problem of shear stress cannot be neglected, there is a need for a higher order shear deformation theory. In this case the third order theory of Reddy [Reddy, 1996] will be used to describe the inner activities.

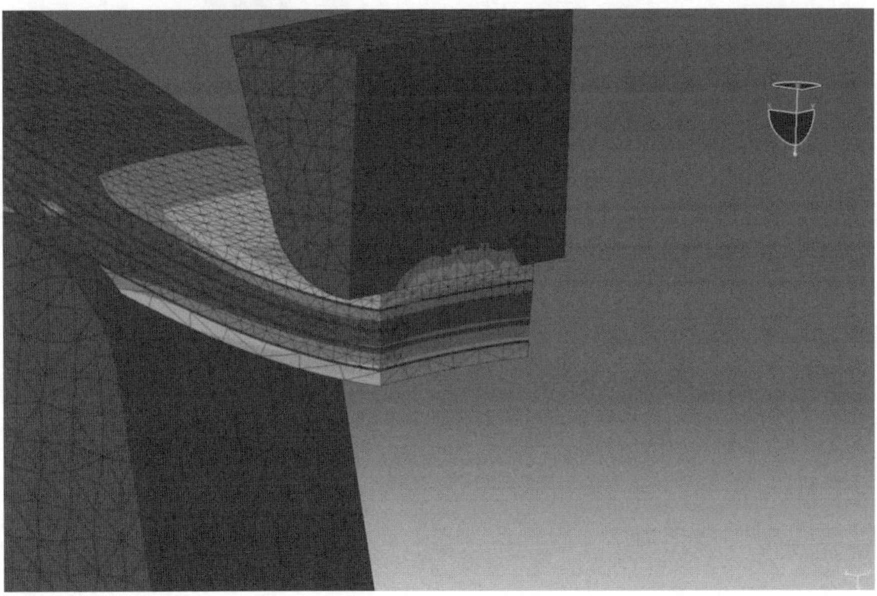

Fig. 6 FE simulation of a 5 layer natural stone laminate resistance.

4.2 Acoustic testing

Besides very good compressive properties the natural stone offers *excellent values of noise insulation.*

In a test in cooperation with the *Fraunhofer Institute* at Stuttgart, sound transmission values were evaluated. Therefore an aluminium sandwich (Fig. **7**) was compared with a natural stone laminate plate of the *same weight*.

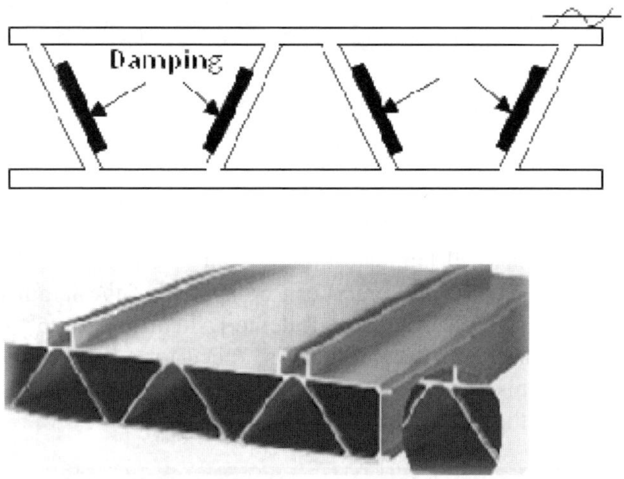

Fig. 7 Aluminum sandwich test specimen for noise insulation.

For the acoustic tests a natural stone called "*Nero Assoluto Zimbabwe*", a magmatic plutonic rock was chosen in an "ABA" composite with aluminium.

With *over 3dB* in the middle, natural stone laminate reaches a *far better insulation level*, than the aluminium sandwich does.

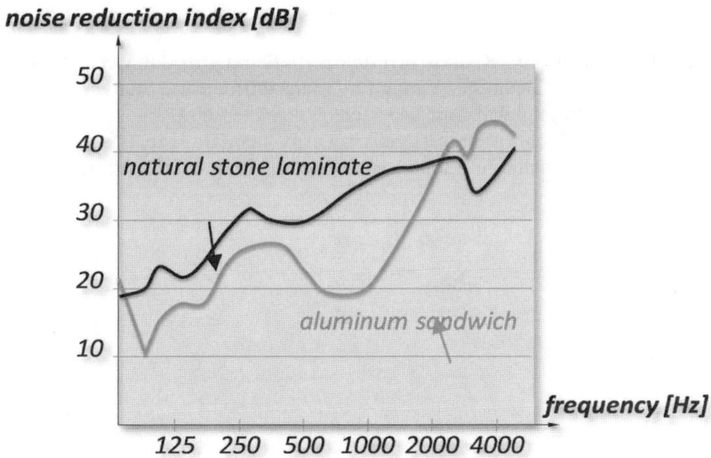

Fig. 8 Noise reduction index of selected testees.

5 Outlook

Next steps are to continue the evaluation of NSL. Therefore the ITD is working together with the University in Ingolstadt which allows the use of very modern laboratories. The ITD has built up a *slow speed impact testing* system for penetrating materials with different projectiles like ice and stone.

Possibilities of the system are:

- Maximum of projectile size 300x300
- Projectile: ice 50 g, stone 10 g
- Distance of acceleration 5 m
- Work pressure 8 bar
- Maximum of velocity 250 m/s

Another point is to determinate substitutes to the used adhesives, e.g. the procedure of *ultrasonic welding*, that shows great results in first tests and even allows operation in high temperature areas. Exceeding investigation must be done in scope of manufacturing *complex shaped structures*.

References

Altenbach H., Altenbach J., Rikards R. (1996), Einführung in die Mechanik der Laminat- und Sandwichtragwerke, Deutscher Verlag für Grundstoffindustrie Stuttgart, Erlangen.
FlightGlobal, www.flightglobal.com, accessed on oct, 19th 2009, http://www.flightglobal.com/assets/getAsset.aspx?ItemID=9116
Reddy J.N.(1996), Mechanics of Laminated Composite Plates, CRC Press, USA.

DELAMINATION OF COMPOSITE STRUCTURES AND FAILURE MODES OF BONDED ELEMENTS

H. Bansemir

Consulting for Composites Light Weight Structures and Knowledge Management
Geitnerweg 4, D-81825 München
Email: horst@bansemir.eu, Tel.: 089/4300968

Abstract Composite structures and bonded elements are widely used in the design of aerospace structures. In 1967 the BO 105, a product of the former helicopter division of MBB, now Eurocopter Germany, flew for the first time with composite rotor blades. Until today a great number of different civil helicopters have been developed. The operative blade consists of a metallic erosion strip, a titanium fitting and a glass and carbon fibre main rotor blade body. Special bonding procedures are used for connecting the metal structures to the blade body. In order to save weight, a composite fuselage of the BK 117 was designed in the 80´s; newer helicopters, such as "Tiger" and NH 90, were designed with composite materials in order to match the military crash requirements. The understanding of delaminations and failure modes of bonded structures were necessary for the substantiation and certification.

1. Introduction

Bonded structures in the airplane industry were already used before the Second World War. Goland and Reissner (1944) analyzed the bonded joints with respect to the shear and transverse stresses. These stresses (including "peeling stresses") were also calculated by Volkersen in 1963. In 1973 Bansemir developed a method for analysing the single lap and double lap joints under various loading conditions. The applied "Transfer Matrix" Method allowed enclosing discontinuous stiffnesses in the analysis. Until today the basic "Shear-Lag" theory is used by the airplane industry to determine the nonlinear shear stresses. The early developments in the 60´s at MBB, now Eurocopter Germany, show an intensive use of composites in the design of rotor blades. More recently "Fracture Mechanics" has been used to determine the strength of tapered composite helicopter spars.

J. Wellnitz et al. (eds.), *Sustainable Automotive Technologies 2010*,
DOI 10.1007/978-3-642-10798-6_21, © Springer-Verlag Berlin Heidelberg 2010

2. Examples for bonded helicopter structures

Light weight structural components are of special importance for the design of helicopters. The use of composite materials is therefore necessary. The rotor blades are designed with the help of glass and carbon fibre composites. These structures are to be bonded with metal components in order to fulfill the customer and certification requirements. The need for sufficient erosion protection leads to a design where the composite blade body is bonded to a metal shell at the leading edge. For the metal shells steel, titanium and nickel are used, depending on the requirements of different helicopters. For the bonding procedure the metal shell is co-cured with the "wet" composite body or separately bonded to the blade. Often, the adhesive FM 73 (Cytec Specialty Chemicals) is used for bonding. For the calculation of the stresses the well known "Shear Lag Theory" is used, see also the referenced literature of the authors Goland and Reissner (1944), Volkersen (1963) and Bansemir (1973).

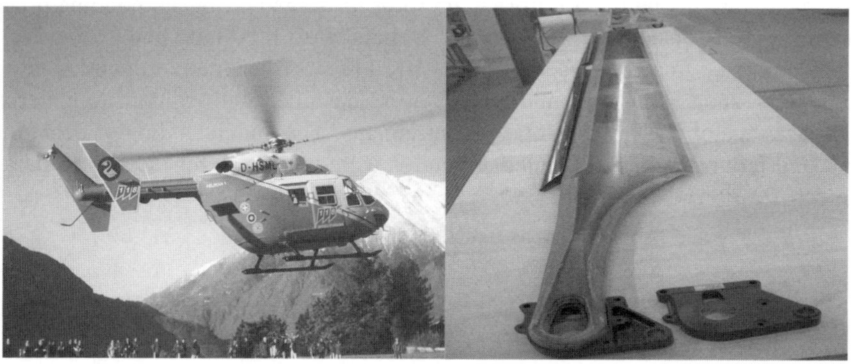

Fig. 1 The commercial BK 117 helicopter with the components erosion strip, titanium fitting and composite blade body.

The composite fuselage of the helicopter BK 117 shown in Figure 2 was developed in the 80´s as a basis for a new generation of helicopters. The composite fuselage with two side panels consisted of many elements bonded together. The frames of the shell were bonded together in different conditions ("co-curing" and "co-bonding"). Other elements, like for example the maintenance footsteps, are bonded in a pre-cured condition to the panel.

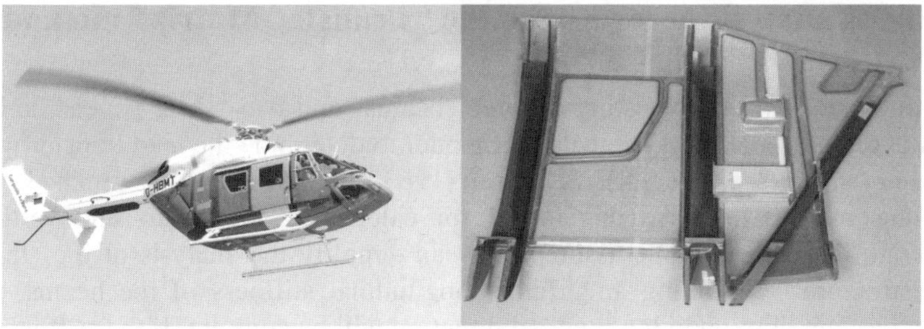

Fig. 2 The BK 117 helicopter with composite fuselage and side wall.

Foil adhesives with and without fabric carriers are used to bond parts with small tolerances. For bonding structures with larger tolerances, high temperature curing one-component adhesives are used. On the other hand, two component adhesives are used to comply with even higher tolerances. Other materials in use are: splicing adhesives, sealants and laminating resins for repair. With the introduction of bonded structures in the helicopter design, environmental influences such as moisture and higher temperatures, as well as the effects of hidden impact damages have to be considered for certification.

The tail rotor ("Fenestron") drive shaft of the EC 135 is an important and complicated structural element. On the one hand the torsion moment has to be transferred, on the other hand the safety margin to the critical rotational speed must be kept. In addition, the certification authorities require the application of the "damage tolerance principle". The fatigue loads together with impact damage corresponding to a load of 25 joule have to be demonstrated by tests. After the dynamic test the residual strength together with higher temperature and moisture has to be demonstrated. The life-time can be calculated with the help of the Miner-Palmgreen rule. The carbon composite drive shaft is bonded and riveted to a titanium fitting. The gluing complies with the rigidity and strength between the different elements and the riveting gives a margin of safety in case of failure of the bonding.

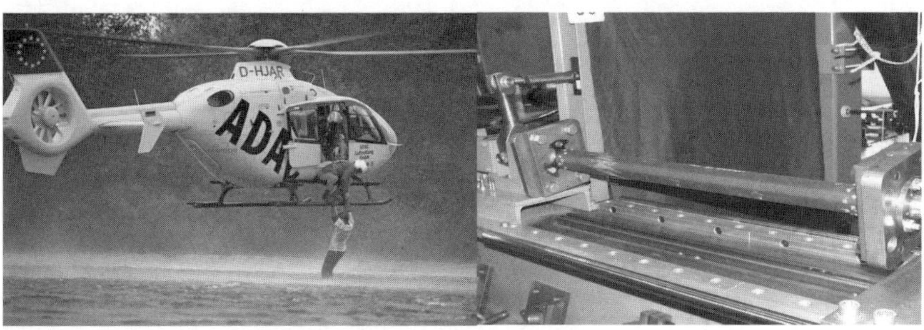

Fig.3 The EC 135 helicopter and the "Fenestron" carbon composite drive shaft.

3. Stress analysis according to the "Transfer-Matrix" method

The structural element consisting of three "beams" and two "shear panels" in between is often used as a basic analytical model to determine the load transfer with the help of shear (see Figure 4). Volkersen (1963), Schnell and Czerwenka (1970) and Bansemir (1973) used this model for calculating the shear- and vertical stresses in the panels. Some assumptions can simplify the analysis of the "three beam structure". Assuming an infinite longitudinal stiffness of the beams, the shear stress distribution in longitudinal direction will be constant. If the stiffness is finite, the so called "Shear Lag Theory" is achieved by taking into account the infinite transverse panel stiffness. With this stiffness the shear stress distribution complies with the edge condition of zero shear stresses. The transverse stresses can be calculated with the help of the equilibrium condition in the panel. The model for the analysis is formulated by Bansemir (1973) with the help of ten differential equations of the first order. The upper left model of Figure 4 shows a double lap joint with the typical loads. The shear and the transverse stresses are shown in Figure 5 for different bending stiffnesses of the upper beam (e.g. very low, stiff and infinite stiff). The shear stresses are zero for the edges x=0 and x=a. The influence of the bending stiffness of the upper beam is low. Moreover, the transverse stresses are linearly distributed across the thickness of the adhesive. Critical for the strength is the tensional stress called also "pealing stress". Similar analyses were performed for the other load cases shown in Figure 4.

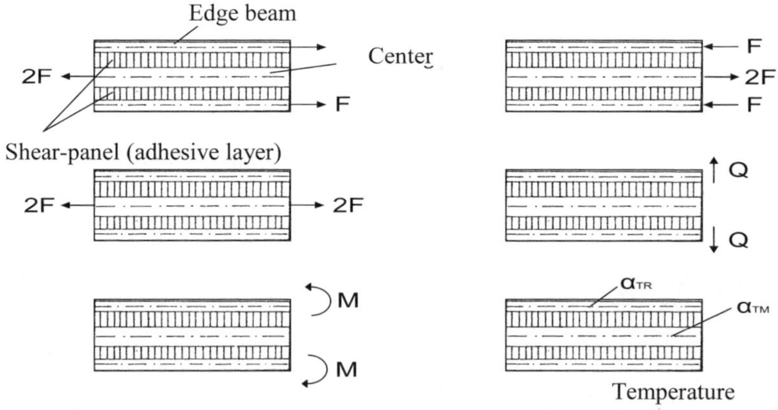

Fig. 4 The "three beam structure" with different loading situations.

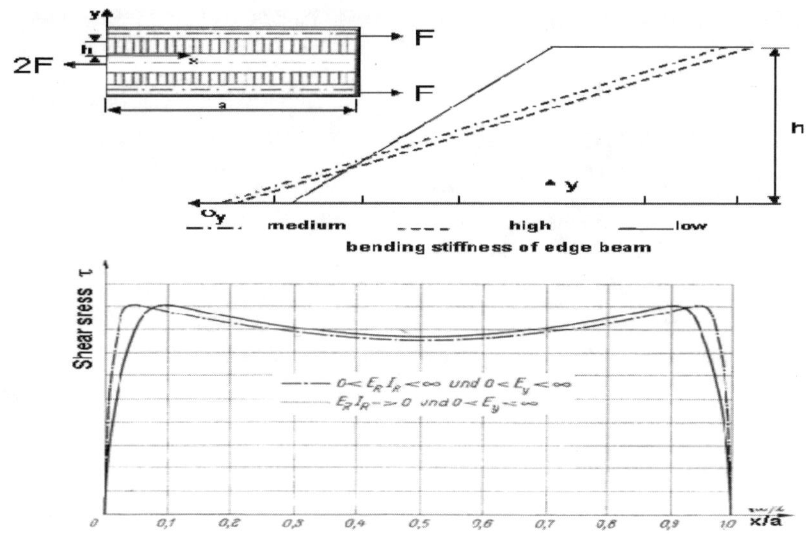

Fig. 5 Shear stress distribution for a double lap joint including the transverse stresses for different bending stiffnesses of the upper beam.

4. Tension and torsion loaded tapered laminates

Advanced rotor blades such as the "Flexbeam" of the EC 135 have tapered distribution of fibre composites. The transfer of longitudinal forces in the reinforced areas of the spar results in peak shear stresses. The "Shear Lag Theory" requires for calculating the transferable stresses the thickness, shear modulus and the shear strength of the adhesive. The formulas obtained by applying the fracture mechanics rules represent the transferred stress as a function of $G_{C,II}$ (critical strain-energy release rate). We can equate the Shear-Lag equation with the fracture mechanics rule and get a relation between the critical strain-energy release rate and the adhesive data. The critical strain-energy release rate $G_{C,II}$ can be measured by simple test methods, e.g. with the Transverse Crack Tension (TCT) specimen. The relation between fracture mechanics and the "Shear-Lag Theory" was derived by Rapp (1993).

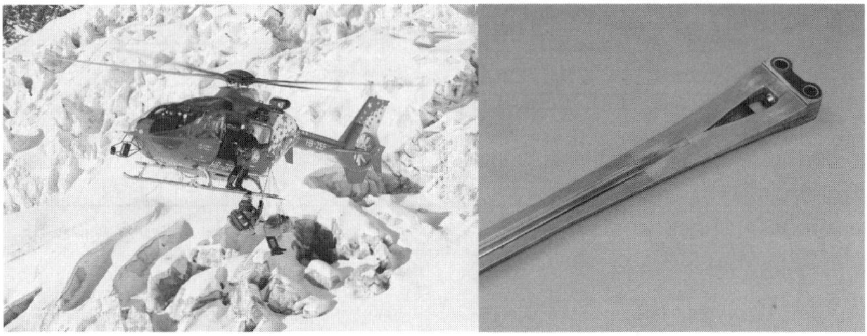

Fig.6 The helicopter EC 135 and the tapered rotor blade "Flexbeam".

5. Summary

The safe design of bonded composite structures is essential for light weight helicopter components. 1967 the BO 105 flew for the first time using composite rotor blades. Metallic erosion strips are bonded to the composite blade body. Analytical methods are presented including an improved "Shear-Lag" theory and the strain energy release rate derived from fracture mechanics.

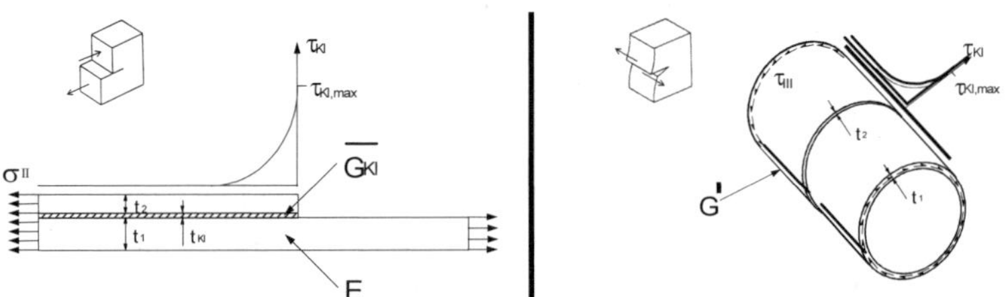

Tension	Torsion
$\sigma_{II} = \tau_{KI,max} \cdot \sqrt{\frac{t_{KI}}{G_{KI}}} \cdot \sqrt{\frac{E \cdot t_1}{t_2 \cdot (t_1 + t_2)}}$	$\tau_{III} = \tau_{KI,max} \cdot \sqrt{\frac{t_{KI}}{G_{KI}}} \cdot \sqrt{\frac{G' \cdot t_1}{t_2 \cdot (t_1 + t_2)}}$
$\sigma_{II} = \sqrt{2 \cdot G_{C,II}} \cdot \sqrt{\frac{E \cdot t_1}{t_2 \cdot (t_1 + t_2)}}$	$\tau_{III} = \sqrt{2 \cdot G_{C,III}} \cdot \sqrt{\frac{G' \cdot t_1}{t_2 \cdot (t_1 + t_2)}}$
$G_{C,II} = \frac{1}{2} \cdot (\tau_{KI,max})^2 \cdot \frac{t_{KI}}{G_{KI}}$	$G_{C,III} = \frac{1}{2} \cdot (\tau_{KI,max})^2 \cdot \frac{t_{KI}}{G_{KI}}$

Fig. 7 Tapered structure loaded by tension and torsion.

Acknowledgments

I would like to thank Eurocopter Deutschland GmbH for providing the Figures 1, 2, 3 and 6.

References

Goland M., Reissner E. (1944) *The stresses in cemented joints* J. Appl. Mech. 11,

Volkersen O.(1963) *Untersuchungen zur Theorie der Klebverbindungen.* Jahrbuch der Wiss. Luft- und Raumfahrt S.299-306.

Bansemir H. (1973) *Krafteinleitung in versteifte orthotrope Scheiben.* Ing. Archiv, Bd. 42, H.2

Bansemir H. (1974) *Spannungsanalyse von abgestuften, überlappten Klebeverbindungen (Step-Lap-Joint).* MBB GmbH TN-D124-10/74

Vater B. (1979) *Theoretische und versuchstechnische Untersuchungen verschiedener Einflüsse auf Klebverbindungen.* Diplomarbeit an der Hochschule der Bundeswehr München

Schnell W., Czerwenka G. (1970) *Einführung in die Rechenmethoden des Leichtbaus II* Mannheim BI Hochschultaschenbücher 125/125a

Rapp H. (1993) *Berechnung von Delaminationen in Aufdoppelungen und Ausschäftungen.* Unpublished Report of Eurocopter Germany GmbH ECD-DE133

Bansemir H., Emmerling, S. (1999) *Fatigue Substantiation and Damage Tolerance Evaluation of Fiber Composite Helicopter Components,* Corfu Greece, Applied Vehicle Technology Panel: Applications of Damage Tolerance Principles for Improved Airworthiness of Rotorcraft

Bansemir H. (2009) *Spannungsverteilungen von ein- und mehrschnittigen Klebverbindungen,* Landshuter Leichtbau Colloquium 28.2.-27.2.2009

Bansemir H. (2009) *Strukturmechanische Aspekte von Klebverbindungen im Hubschrauberbau,* 9. Kooperationsforum, Industriekreis Kleben im Automobilbau, Nürnberger Akademie, Gewerbemuseumsplatz 2, 90403 Nürnberg, 28.4.2009

APPLICATION OF SUPERSONIC FLAME SPRAYING FOR NEXT GENERATION CYLINDER LINER COATINGS

A. Manzat, A. Killinger, R. Gadow

Institute for Manufacturing Technologies of Ceramic Components and Composites (IFKB), University of Stuttgart, Allmandring 7b, 70569 Stuttgart, GERMANY,
E-mail: andrei.manzat@ifkb.uni-stuttgart.de, ifkb@ifkb.uni-stuttgart.de

Abstract Rising demands for ecologically friendly automotive engines require a significant decrease in fuel consumption and emissions. Tribologically functional coatings applied by supersonic flame spraying help in boosting the engine efficiency by reducing the internal friction and improving the durability and wear resistance of the cylinder running surface, which facilitates engine downsizing. Also, the high corrosion resistance enables the use of bio fuels. In addition, the tailored surface topography of the thermal spray coatings shows the benefit of reducing the oil consumption resulting in lower polluting emissions. The thermally sprayed coatings were applied using HVOF and HVSFS processes together with a specially designed spray gun trajectory in order to achieve a fast and cost efficient coating procedure. Several different coating materials, including novel nanostructured powders, have been investigated and compared to state-of-the-art cylinder liners.

1. Introduction

One major approach for low emissions and reduced fuel consumption is the deployment of complex and expensive AlSi-alloys showing advanced structural stability, but on the other hand increase the casting efforts.

An approach to solving these problems is the application of thermally sprayed coatings on inferior but easy to process Al-alloy for engine blocks and thereby increasing their aptitude for a use as lightweight crankcases with improved tribological properties. Most of the approaches so far include the use of specially adapted, small scale torches, e.g. internal rotating spray torches based on Atmospheric Plasma Spraying (APS), Wire Arc or Plasma Transfer Wire Arc (PTWA) techniques (Barbezat *et al.* 1998; Bobzin *et al.* 2008).

In this research a standard HVOF torch has been used, which facilitates the application of high quality coatings and also overcomes the limitations in feed stock choice. Contrary to internal rotating torches, the HVOF torch is operated outside the cylinder bore with the cylinder liner rotating (Buchmann *et al.* 2002).

J. Wellnitz et al. (eds.), *Sustainable Automotive Technologies 2010*,
DOI 10.1007/978-3-642-10798-6_22, © Springer-Verlag Berlin Heidelberg 2010

2. Experimental

The coatings were deposited on aluminum cylinder liners and inline-4-cylinder engines representing the typical size of passenger car engines (82 mm bore, 32,500 mm^2 coated surface) and gray cast iron liners for large commercial vehicle engines (131 mm bore, 82300 mm^2 coated surface). Deployed coating materials are shown in Table 1.

All liners have been grit-blasted prior to the coating process. The coatings have been applied using a GTV HVOF-torch "Top Gun G" with propane as combustion fuel. TiO_2/TiC-coatings have been handled in a suspension due to the small particle size and therefore been applied through suspension flame spraying (HVSFS). A modified HVOF "Top Gun G" torch has been used for this application (Killinger *et al.* 2006).

Table 1 Deployed coating materials.

Brand Name	Manufacturer	Composition	Particle Size (d,50)
Nano3	Durum	Fe-alloy	25 µm
Diamalloy 1008	Sulzer-Metco	FeCrMo	36 µm
Infralloy 7412	Inframat	WC/Co	31 µm
P25	Evonik	TiO_2	0.02 µm
STD 120	H.C. Starck	TiC	2 µm

All internal coatings have been deposited with the torch being manipulated by a computer-controlled 6-axis robot. The trajectory of the torch can be described as an ellipsoidal movement to account for the geometrical constraints imposed by the process of internal coating with an externally mounted torch and the need of a constant spraying distance and constant step size on the coated surface in order to realize homogeneous properties of the coatings.

Since the deposition efficiency is depending on the spraying angle and the thermal conditions, elaborate adaptations of the torch kinematics are necessary for being able to achieve a constant coating thickness. The main adjustments regard the speed of the torch movement along the predetermined trajectory as the speed of transition over a surface also influences the resulting coating thickness. This generally results in a regular coating thickness with deviations as low as ± 6 %, meaning below ± 20 µm for a 300 µm coating, depending on the deployed coating material.

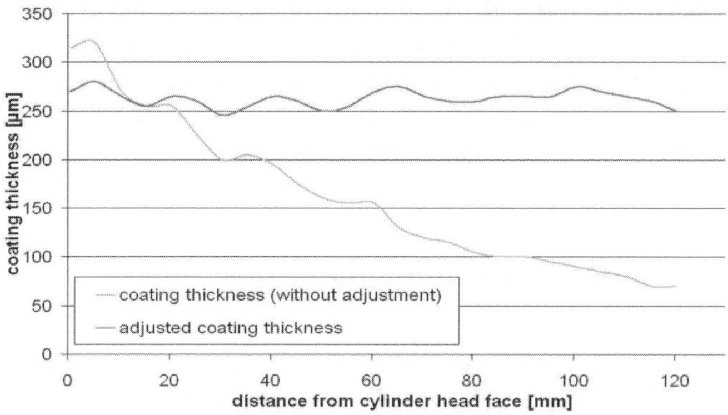

Fig. 1 Exemplary coating thickness gradient before and after speed adjustment.

This adaptation also resulted in significantly lower processing times as there was no need for additional allowance to account for inhomogeneity in the coating and also reduces the costs for mechanical treatments, e.g. lower stock removal during honing.

Another geometrical requirement affected the upper ending of the coating. Due to the limited resistance against possible lateral stresses applied by the cylinder head, the coating has to run out at least one mm under the cylinder head face to ensure a non-contacting operation during the engine run. This issue has been solved by using a specially developed masking, which led to a gradually decreasing coating thickness with a smooth edge on the upper coating end.

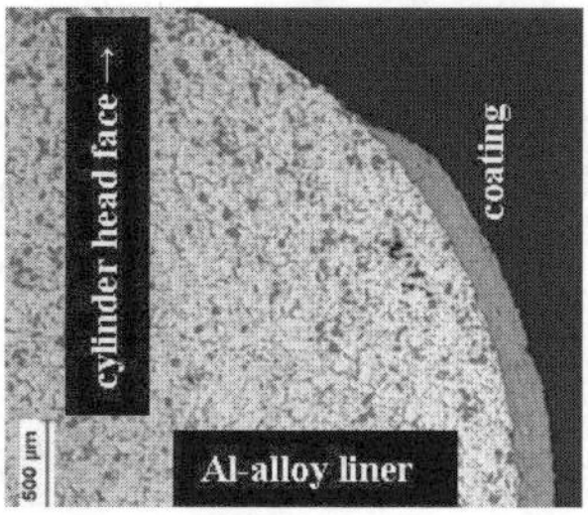

Fig. 2 Upper coating ending (Fe-alloy on Al liner).

The coating has been applied by means of several torch transitions with a deposition of up to 74 µm per transition depending on coating powder, resulting in an average coating thickness of 300 µm respectively 200 µm for TiO_2/TiC.

Despite the disadvantageous spraying angles, acceptable overall deposition efficiency could be achieved, especially for the smaller aluminum liners, the deposition efficiency reached more than 55 %.

All trajectories and speed adaptation applied during liner coating can be transferred to engine block coating without any modification leading to the same coating thickness shapes. After being coated, the liners have been honed with the same crossing groove structure like the state-of-the-art gray cast and Al-alloy liners, except that the grooves could be formed significantly shallower. The pores on the surface form micro cavities which improve the oil retention abilities superseding the deep honing grooves. This results in a smooth surface structure with small pores which are typical for thermally sprayed coatings (Flores *et al.* 2007).

3. Material analysis and characterization

The material screening was focused on the development of high-efficiency coating systems for tribological applications with regard to low friction and wear coefficients and high corrosion resistance as well as enhanced operational performance.

Hardness: The micro hardness has been determined with a micro-indenter from Fischer, calculating a HV0.1 micro hardness according to DIN 50359 / ISO 14577 tested on polished cross sections of the samples.

Porosity: The porosity of the applied coatings has been determined by digital image analysis of cross section micrographs.

Table 2 Porosity and micro hardness.

	Fe-alloy	FeCrMo	Cr_3C_2/NiCr	NiCrBSi	TiO_2/TiC	WC/Co
av. porosity [%]	1.1	2.9	1.2	1.2	1.4	0.1
$HV_{0.1}$	810	710	1000	775	800	1330

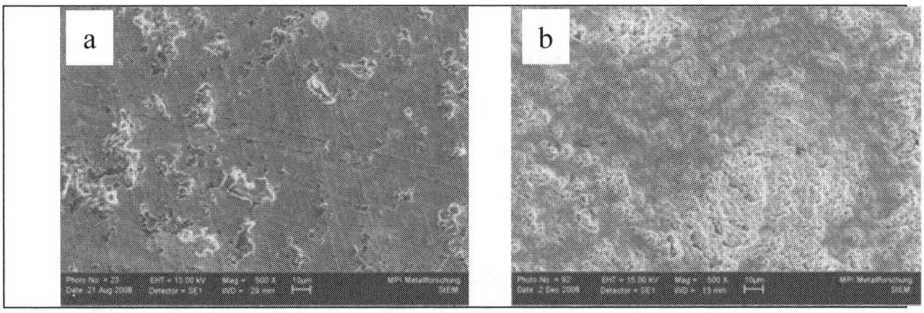

Fig. 3 Exemplary SEM-images of honed surfaces of (a) Cr_3C_2/NiCr and (b) TiO_2/TiC.

4. Tribological analysis

Ball-on-disk tests have been performed with both coated and non-coated liners as a comparison to state-of-the-art gray cast and Al-alloy liners. All tests have been conducted on honed surfaces and under ambient conditions (22 °C).

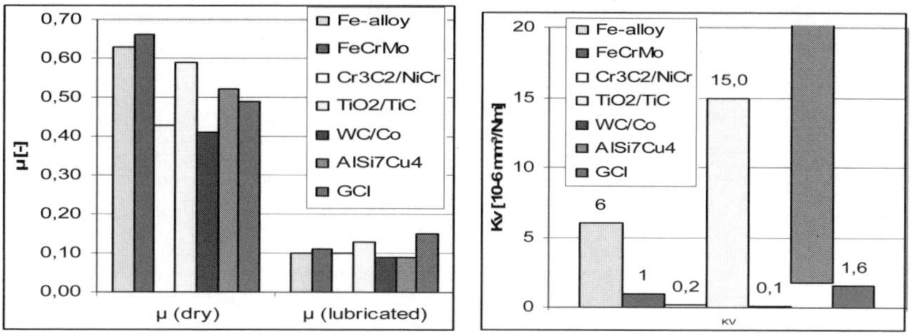

Fig. 4 Volumetric wear coefficients (K_V) and friction coefficients of the coatings.

The friction coefficient has been measured under non-lubricated conditions with an alumina ball as counterpart, which accentuates the wear and load on the tested sample. The friction tests have been performed during 100,000 cycles at 10 N load with oscillating motion of 5 mm displacement at a speed of 70 mm/s. After the friction tests the wear volume has been determined and the wear coefficient has been calculated according to Equation 1 (Lopez 2008),

$$K_V = \frac{V_e}{F_N \cdot s} \tag{1}$$

where K_V represents the volumetric wear coefficient [mm³/Nm], V_e the wear volume [mm³], F_N the normal load force [N] and s the sliding distance [m].

In addition a fretting test has been performed at increasing load starting at 10 N and of 55 N. The speed was 70 mmps with an oscillating motion of 5 mm displacement. A 100Cr6 ball (5 mm diameter) has been used as counterpart for the samples because of its higher fretting tendency leading to shorter failure times. For these tests, a poor lubrication environment has been simulated. The lubricant (motor oil 15W-40) has been applied once at the beginning of the test to the 100Cr6 ball.

The gray cast iron sample showed an increasing friction coefficient after approximately 100 cycles, which is the indicator for incipient failure. The AlSi17Cu4 sample resisted for 1,250 cycles before showing signs of failure. These two samples are state-of-the-art and are being used in production passenger car engines. The FeCrMo-coated sample showed a constant friction coefficient with a light increase until failure. The TiO_2/TiC coated sample showed a constant but rather high friction coefficient before fretting occured. Outstanding performance has been achieved with the WC/Co coating which lasted for 150,000 cycles having a very low friction coefficient.

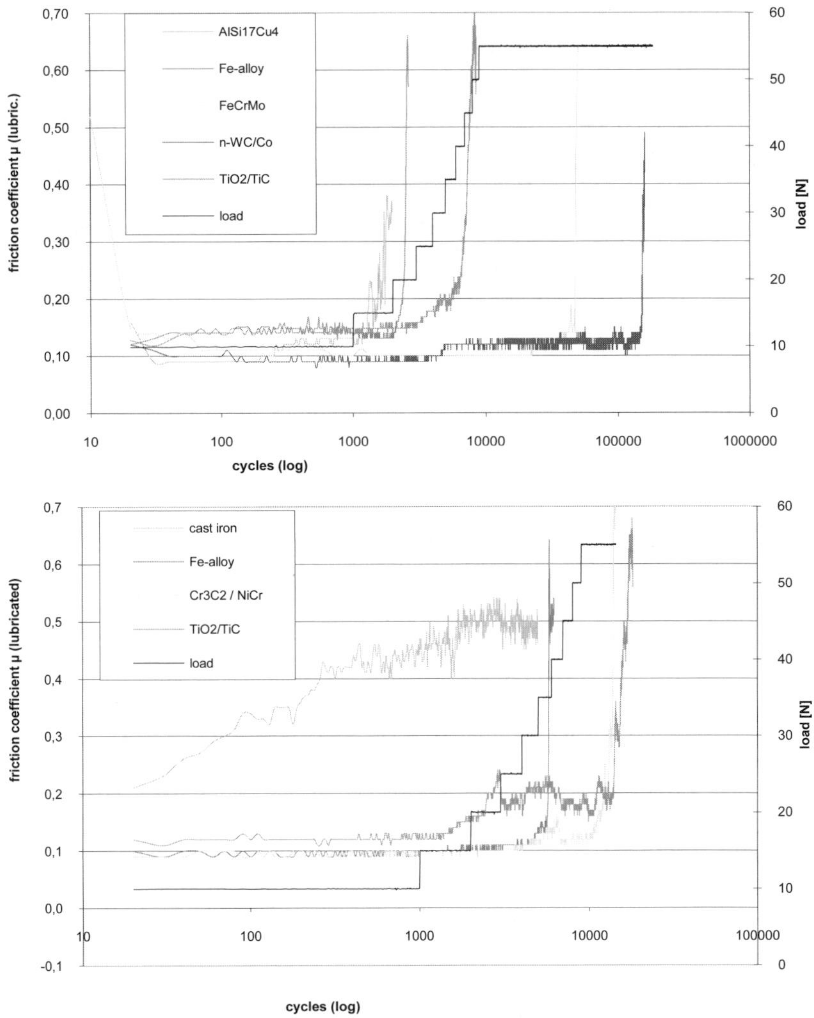

Fig. 5 Friction coefficient (fretting test); coatings on Al (top) and on gray cast iron (bottom).

For the gray cast liners a similar result has been achieved from the fretting tests. The non-coated gray cast sample showed an instantly increasing friction coefficient resulting in failure. The Fe-alloy sample lasted for 4,300 cycles before incipient failure. The Cr_3C_2/NiCr coating showed a constant coefficient for 5,000 cycles with a slight increase until failure commenced. A similar behavior could be observed for the TiO_2/TiC coating except for the higher friction coefficient throughout the fretting. The noticeable difference between the TiO_2/TiC coatings on Al-Alloy liners and gray cast iron liners result from the different liner material hardness and the different coating conditions due to the varying geometrical properties.

5. Conclusions and summary

The only structures on the gray-cast and Al-alloy samples are the deep honing grooves, whereas during the honing process the pore structure of the spray coatings become exposed and result in locally confined non-adjunctive micro cavities on the surface. These micro cavities aid in holding oil and therefore form micro pressure chambers supporting the hydrodynamic state of friction reducing wear and the friction coefficient. This especially helps during friction with low relative motion speeds between liner surface and piston ring occurring near the reverse points. In addition the micro pressure chambers reduce the tendency of fretting in deficient lubrication states of the tribological system, e.g. driving with low oil level or during off-road driving. In addition, a significant improvement in the durability of the engines can be expected according to the very low wear coefficients of the applied coatings. By deployment of the specially adjusted flame spray torch kinematics the layer deposition can effectively be controlled, which results in lower processing times during coating and reduces after-treatment efforts.

Tribological behavior of metallic and ceramic surfaces in internal combustion engines can be improved by protective coatings. Tribological investigations show low friction coefficients and an excellent wear resistance under deficient lubrication conditions. HVSFS was successfully used to apply fine structured, tribologically functional coatings on aluminum and gray cast cylinder liners. This new approach offers the possibility of selecting new coating materials and compositions (Gadow *et al.*, 2008).

References

Barbezat G., Keller S. and Wuest G. (1998) *Internal plasma spray process for cylinder bores in automotive industry*, Proceedings of the 15th International Thermal Spray Conference, Nice

Bobzin K., Ernst F., Zwick J., Schlaefer T., Cook D., Nassenstein K., Schwenk A., Schreiber F., Wenz T., Flores G. and Hahn M. (2008) *Coating Bores of Light Metal Engine Blocks with a Nanocomposite Material using the Plasma Transferred Wire Arc*, Journal of Thermal Spray Technology

Buchmann M., Gadow R., Killinger A. and López D. (2002) *Verfahren und Vorrichtung zur Innenbeschichtung von Hohlräumen durch thermisches Spritzen,* German Patent: DE 102 30 847

Flores G., Hoffmeister H.-W. and Schnell C. (2007) *Vorbehandlung und Honen thermischer Spritzschichten*, Jahrbuch Schleifen, Honen, Läppen und Polieren

Gadow R., Killinger A., and Rauch J. (2008) *New results in High Velocity Suspension Flame Spraying (HVSFS)*, Surface & Coatings Technology

Killinger A., Kuhn M. and Gadow R. (2006) *High-Velocity Suspension Flame Spraying (HVSFS), a new approach for spraying nanoparticles with hypersonic speed*, Surface & Coatings Technology

Lopez D. (2008) *Übeschallflammspritzen (HVOF) von metallurgischen und cermetischen Schichten für Zylinderlaufflächen*, Ph.D. thesis, Shaker Verlag

ENHANCED SHAPE MEMORY ALLOY ACTUATORS

M. Leary, J. Mac, M. Mazur, F. Schiavone, A. Subic

RMIT University, Bundoora, Victoria 3083, Australia
Phone: +61 3 9925 6177, E-mail: martin.leary@rmit.edu.au

Abstract Shape memory alloys (SMA) provide an opportunity to develop ultra-light mechanical actuators with far superior economic and environmental performance to traditional actuators. Despite their many favorable attributes, SMA actuators are subject to a series of inherent limitations, in particular, a relatively slow expansion rate. The authors have proposed a novel method that enables a significant enhancement of the heat transfer rate in a cost effective and robust manner. An advanced predictive model has been developed to enable the optimisation of a virtual prototype before engaging in experimental validation and commercialisation of this novel technology.

1. Introduction

Shape memory alloys (SMA) provide an opportunity to achieve significant environmental benefit over traditional actuators. SMA actuators are subject to a series of constraints to implementation, in particular, a relatively slow expansion rate, which typically occurs by free convective heat transfer to the ambient environment (Van Humbeeck 1999; Huang 2002). Alternate cooling methods have been proposed, including forced convection with a highly conductive media (Rediniotis and Lagoudas, 2002). This method allows very high actuation frequencies, however the requisite hardware introduces a significant cost and weight penalty.

The authors have proposed a novel method that enables an increased heat transfer rate in a cost effective and robust manner (Leary, Schiavone and Subic 2009). The benefits of the proposed method have been proven experimentally by the authors for a particular specimen configuration. In preparation for commercialisation of this technology, an advanced predictive model is required to enable the design refinement and optimisation of a virtual prototype before engaging in experimental validation.

This work reports on a Finite Difference Equation (FDE) implementation to enable greater predictive certainty in the specification of lagging parameters. An assessment of FDE attributes according to the design requirements identified the preferred FDE implementation. The adopted implementation was developed to enable rapid simulation of SMA parameters prior to physical testing. These out-

J. Wellnitz et al. (eds.), *Sustainable Automotive Technologies 2010*,
DOI 10.1007/978-3-642-10798-6_23,

comes provide a template for the implementation of SMA actuators that will enable significant mass reduction and increased design flexibility in electro-mechanical automotive systems.

2. The design problem

Shape memory alloys undergo a crystallographic transformation in the solid state. The transformation effect results in a significant volumetric change which can be utilised in the design of ultra-light mechanical actuators.

Typically, response time is a critical constraint to the implementation of SMA actuators in automotive systems. SMA actuation is achieved by resistive heating which is acceptably fast. However, SMA cooling is achieved by heat transfer mechanisms that are inherently slower than resistive heating (Van Humbeeck 1999), and response time is dominated by the associated heat transfer rate (Leary and Schiavone 2008).

A novel method has been developed by the authors to enable an increased rate of convective heat transfer by the addition of a lagging material, enabling optimisation of the heat transfer rate in a cost effective and robust manner (Leary, Schiavone and Subic 2009). The method utilizes cylindrical heat transfer media, i.e. lagging, to manipulate the heat transfer rate (Figure 1).

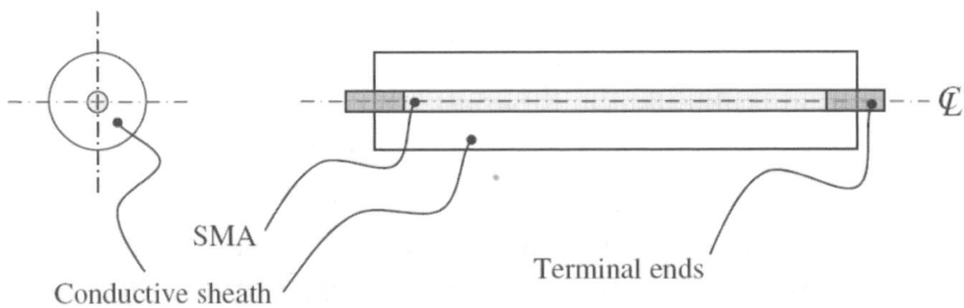

Fig. 1 Novel method to manipulate the heat transfer rate, after (Leary, Schiavone, Subic 2009).

Increasing lagging thickness results in an *increased* resistance to conductive heat transfer concurrently with a *reduced* resistance to convective heat transfer. The total resistance to heat transfer displays a local minima, known as the *critical radius*, r_{cr} (Holman 1990). Cylindrical lagging allows heat transfer resistance to be either increased or decreased depending on whether the lagging is greater-than or less-than the associated critical radius. Selection of the lagging media and dimensions allows control of the critical radius to optimise the cooling rate within the design constraints.

3. Finite Difference Equations

The Second Law of Thermodynamics identifies that a temperature gradient will result in a transfer of heat. The associated local temperature is predicted by the one-dimensional heat diffusion equation (Equation 1).

$$\frac{\partial T(x,t)}{\partial t} = \alpha \frac{d^2 T(x,t)}{\partial x^2} \tag{1}$$

A Finite Difference Equation (FDE) provides an approximate solution to differential equations such as the one-dimensional heat diffusion equation, and was applied in this work to enable robust and rapid estimation of the influence of lagging parameters on SMA response. Numerous FDE embodiments are available, each with associated truncation error and stability.

3.1 Truncation error

The intent of the FDE approximation is to replicate the associated differential equation. However, the discretization of the problem results in a difference between these solutions, known as the truncation error. Truncation error is dependant on the associated FDE method, and can be reduced by increasing mesh size at the expense of computational efficiency.

3.2 Stability

The FDE method originates from a Taylor series expansion, and consequently there is a possibility for instability in results. Stability is a function of the FDE embodiment and can be evaluated by completing a Von Neumann Stability Analysis (Richtmyer and Morton 1957).

Of the proposed FDE methods the following have been found to be particularly effective in solving the heat transfer problem:

1. Forward difference equation (Explicit)
2. Backward difference equation (Implicit)
3. Central difference equation (Crank and Nicholson)

3.3 Forward difference equation (Explicit)

The forward difference equation uses known parameters at an initial time step to calculate values at the next time step. For example, the explicit method considers the temperature of three connected nodes, T_{m-1}, T_m, T_{m+1}, at time, p, and uses their known values to calculate the temperature of the central node, T_m, at time $p+1$ (Equation 2). This method is computationally straightforward as each equation has only one unknown, and it can be solved without simultaneous methods. However, the forward difference equation is unstable if the associated Fourier number ($Fo = \alpha\Delta t / \Delta x^2$) is not maintained between 0 and ½. The associated truncation error is $O[\Delta t, \Delta x^2]$.

$$T_n^{p+1} = FoT_{n-1}^p + (1-2Fo)T_n^p + FoT_{n+1}^p \tag{2}$$

3.4 Backward difference equation (Implicit)

The implicit method of the finite difference equation requires that the initial state of the system is fully understood. From this data an expression for the temperature of nodes in the following time step is created (Equation 3). The expression must be solved simultaneously by the inversion matrix method and requires significantly higher computational resources than the explicit method. However, the associated truncation error is reduced and is independent of mesh selection: $O[\Delta t, \Delta x^2]$.

$$-FoT_{n-1}^{p+1} + (1+2Fo)T_n^{p+1} - FoT_{n+1}^{p+1} = T_n^p \tag{3}$$

3.5 Central difference equation (Crank-Nicholson)

The Crank-Nicolson equation combines the explicit and implicit methods, resulting in an average of the two, yielding a smaller truncation error due to a decreased sensitivity to changes within x. The Crank-Nicolson method is as an implicit scheme in which a series of simultaneous equations are solved for each time step (Equation 4). The Crank-Nicolson method considers six nodes whereas only four nodes were considered in the simple explicit and implicit methods. The Crank-Nicolson method provides high accuracy results and is unconditionally stable, allowing for maximum flexibility in the choice of discretization. (Crank and Nicolson 1947). The truncation error is second order accurate with both space and time, $O[\Delta t^2, \Delta x^2]$, illustrating the minimal effect the discretization of the equation has on the FDE solution.

$$-FoT_{n-1}^{p+1} + (2 + 2Fo)T_n^{p+1} - FoT_{n+1}^{p+1} = FoT_{n-1}^{p} + (2 - 2Fo)T_n^{p} + FoT_{n+1}^{p} \quad (4)$$

4. Simulation

The Crank-Nicholson method was applied to solve the one-dimensional heat equation for a simple polar geometry consisting of a 0.254mm diameter SMA wire surrounded by an annulus of lagging material (Table 1). To simulate a typical actuation scenario, two-stages of boundary conditions were applied to simulate wire contraction (heating) and expansion (cooling) respectively.

Table 1 Material properties applied (Leary *et al.* 2009).

Material	T_i [°C]	ρ [g/cm^3]	k [W/mK]	h [W/m^2]	D_{min} (mm)	D_{max} (mm)
SMA	100	6.45	21.9	–	0.254	0.254
Lagging	21	2.1	0.59	–	0.254	50
Shell	21	0.910	0.12	–	0.254	50
Air	21	–	–	6	–	–

SMA actuation was simulated by applying an initial temperature boundary condition, T_i, associated with the SMA actuation temperature and convection to the environment to a system that was initially at ambient temperature (Table 1). The SMA was assumed to reach its actuation temperature instantaneously. The lagging material temperature increases monotonically, until the heat flow rate stabilises and the system reaches steady-state (Figure 2). Once steady-state conditions were achieved, the SMA wire temperature boundary condition was removed to simulate deactivation. In this stage the SMA wire temperature is determined by cooling to the ambient environment (Figure 3). The exothermic heat transfer of phase change was ignored (Van Humbeeck 1999).

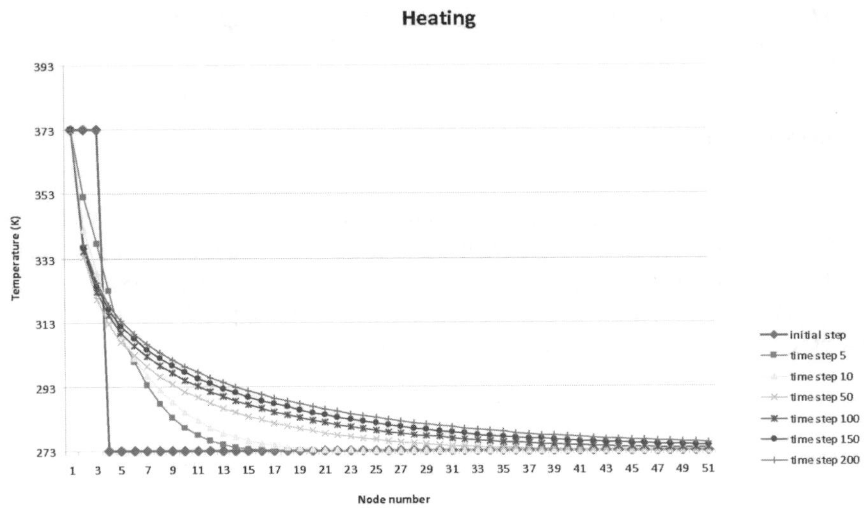

Fig. 2 Representative contraction (heating) phase FDE solution.

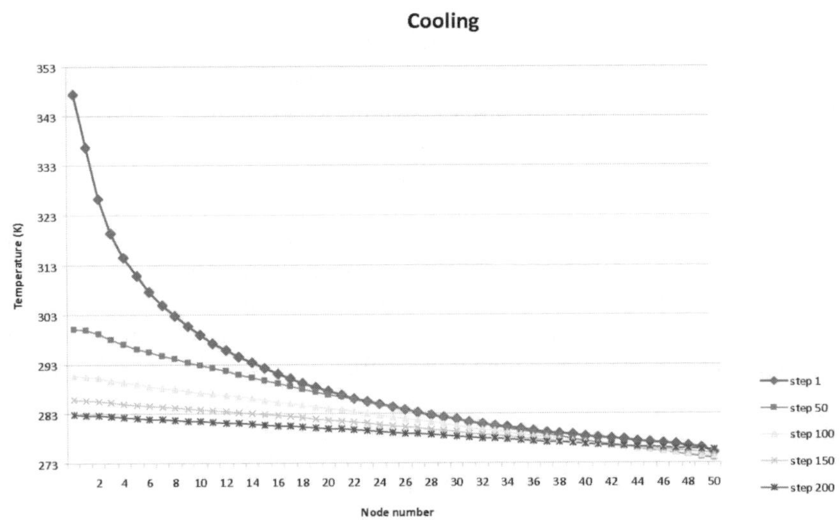

Fig. 3 Representative expansion (cooling) phase FDE solution.

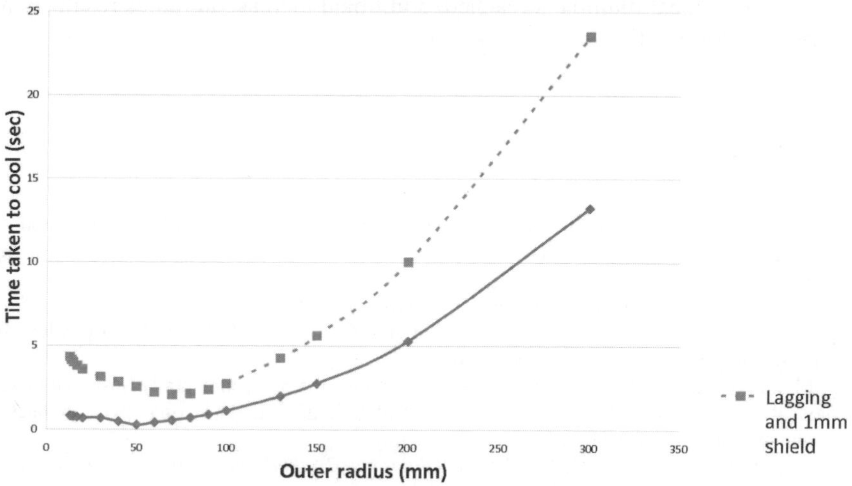

Fig. 4 Representative expansion (cooling) phase FDE solution.

5. Outcomes

The proposed FDE analysis enables rapid simulation of the performance of a range of SMA lagging parameters, providing an opportunity to predict the performance of feasible design parameters before building physical prototypes.

The proposed FDE analysis enables the evaluation of a specific configuration. To assess a larger design space requires an automated Design of Experiments (DoE) approach. Taguchi methods may be applied to generate a partial factorial DoE approach to enable automated analysis of a large number of feasible design configurations. To ensure robust outcomes of the automated analysis it is necessary to ensure that the steady-state condition is reached during the heating phase. This can be achieved by ensuring that the root-mean-square (RMS) temperature difference between last two iterations of the FDE is sufficiently small. The system performance is then characterised by the total time required for the SMA to reach the deactivation temperature (Table 1).

The contraction and expansion phases of the SMA simulation were linked for a range of feasible lagging parameters and the deactivation time recorded. Initial outcomes indicate that:

1. Deactivation time is minimized if paste and shield conductivity are maximized. The design constraints restrict the material selections available for the shell, and the associated conductivity is likely to be significantly lower than the paste (Table 1). This restriction may be mitigated by minimizing shell thickness.
2. The predicted cooling time is compatible with experimental observations and critical radius theory and indicates a reduction of cooling time with increasing

radius until a local minima is reached; additional increase in radius results in an increasing cooling time.

References

Crank, J. and Nicolson, P. (1947). A practical Method for Numerical Evaluation of Solutions of Partial Differential Equations of Heat-Conduction Type, *Proc. Cambridge Philos.*, vol.43, pp. 50-67.

Holman, J.P., (1990) *Heat Transfer, Volume 7.* McGraw-Hill, New York.

Leary, M., F. Schiavone and A. Subic. (2009) Lagging for Control of Shape Memory Alloy Actuator Response Time, *Materials and Design*, in Press.

Leary, M. and Schiavone, F. (2008) Shape Memory Alloys for Environmentally Sustainable Mechanical Actuation. In Subic, A., Leary, M., and Wellnitz, J. (Eds.) Meeting the Challenges to Sustainable Mobility, RMIT University, Melbourne.

Richtmyre, R.D. and Morton, K.W., (1957), *Difference Methods For Initial-Value Problems*, Second Edition, John Wiley & Sons, Inc, United States of America.

Van Humbeeck, J., (1999) Non-medical applications of shape memory alloys. Materials Science and Engineering A, 1999. 273-275: p. 134-148.

DAMAGE BEHAVIOR OF BONDED STRUCTURES UNDER STATIC AND CYCLIC LOADING CONDITIONS

H.-J. Gudladt, St. Froemmel

Institute for Materials Science, Universität der Bundeswehr München,
Werner-Heisenberg-Weg 39, 85577 Neubiberg, Germany,
E-mail: hans-joachim.gudladt@unibw.de, stephan.froemmel@unibw.de

Abstract The effects of bubbles in an adhesive caused by different kind of mixing methods as well as the surface treatment like phosphor acid anodization and the laser treatment were investigated in the light of the mechanical behaviour of a bonded structure. Furthermore, the fatigue behaviour of bonded specimens has been analyzed by using damage quantities like resonance frequency, changing of stiffness and engine power output.

1. Introduction

1.1 Motivation

To improve mobility in a global world, the transport problem can be solved by using more and more light-weight structures for vehicles like cars, air-planes and trains as well as ships. Consequently, bonded structures composed of metal, polymer, glass and / or ceramic have to be used. To understand the mechanical behaviour of these structures, besides FE calculations, the mechanical testing is unambiguously necessary. Up to now, most of the structures have been tested under static loading conditions. To qualify the mechanical behaviour of bonded structures, lap shear tests can be used and, besides the shear stress of rupture, R_m, the kind of fracture (cohesive or adhesive) represents a quantity to define safe structures. In addition, further parameters like the adherent material, the methods of joint preparation and the surface treatment of the substrate as well as the adhesive used may influence the mechanical behaviour.

J. Wellnitz et al. (eds.), *Sustainable Automotive Technologies 2010*,
DOI 10.1007/978-3-642-10798-6_24,

1.2 Theoretical approach

In case of static tests it is relatively easy to define the damage quantity R_m. On the other hand, under real test conditions for cars and air-planes, the cyclic deformation takes the main contribution to the damage behaviour of bonded structures. Now, besides the total failure, a new damage quantity that is sensitive to the cyclic deformation has to be found. Normally, the recording of the S-N (Woehler) curve is the common used procedure to gain information of the cyclic lifetime. Unfortunately, for a given stress amplitude one obtains only the cycle number of the specimen failure, N_f, as a damage information.

For detailed lifetime predictions, a damage quantity that depends on the cyclic deformation history has to be used to qualify the fatigue tests undertaken with bonded specimens. The latter indicates that the damage evolution can be measured cycle by cycle, e.g. in a continuously manner. In the present case, the quantity that represents the amount of damage reflects in change of stiffness. The latter has been measured via the resonance frequency of the fatigue resonance testing machine. The safety of the bonded structure will be discussed in the light of the damage induced by monotonic or cyclic loading conditions.

2. Experimental section

At first, a two component (2k) epoxy adhesive system was investigated to study the influence of the mixing method. To qualify the mechanical behaviour of the bonded structures single-overlap-beams depending on DIN 53281 were used. These samples have been prepared with different surface treatment methods (phosphoric acid anodization; NdYAG-Laser). Thereafter, they were tested under static and cyclic loading conditions.

Substrate specimens of the aluminium alloy AA6009-T6 with 2mm thickness were used. The mechanical properties are shown in table 1.

Table 1 Mechanical properties of the aluminium alloy

	AA6009-T6
UTS	322 MPa
$R_{p0,2}$	286 MPa
E	67200 MPa

To obtain an optimal surface treatment some of the samples were phosphoric acid anodized depending on ASTM D1002 (Figure 1, line 2). In addition, further sample surfaces were treated by a Trumpf marking laser with 1062/1064nm laser wavelength. Parameters of the laser beam conditions are given in table 2. The power of the laser beam was varied between 0% (Figure 1, line 1) and 100% (Figure 1 line 4).

Table 2 Parameters for laser treatment

Frequency	20000Hz	defocus	0µm
Speed	400mm/s	focus diameter	50µm
pulse range	4.00µs	line distance	50µm
pulses / point	1	laser power	0% - 100%

The substrate surface after anodization is just a little bit rougher than the blank one but there exists a strong aluminium oxide coating, as it is well known (Davis, 1993). The laser treatment also produces a kind of aluminium oxide coating which consists of a special macroscopic topography produced by the high energy of the laser beam. The surface material is melted and immediately cooled down so that the molten structure remains on the substrate surface. As high resolution SEM images have shown, laser treatment causes a rough surface structure on a macroscopic, microscopic and also a nanoscopic level.

The effect of the laser treatment on the surface structure is shown in Figure 1 (lines 3 & 4). As a consequence, its influence on the mechanical behaviour of the bonded structure is presented in Figure 2. The shear strength depends on the power of the laser beam and, consequently, on the degree of surface roughness. As it is shown, a laser beam power lower than 50% does not improve the resulting shear strength. A power higher than 60%, however, leads to significant better results of the shear strength.

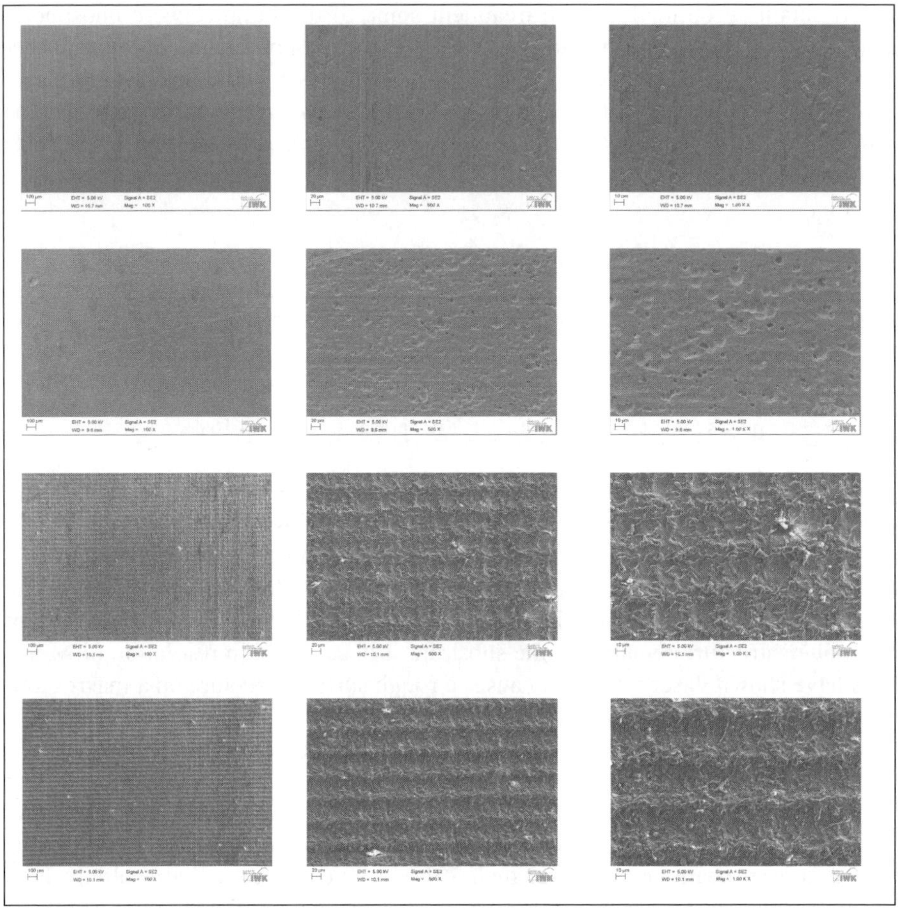

Fig. 1 SEM-images of a blank (line 1) and an anodized (line 2) aluminium surface. The micrographs in lines 3 and 4 represent the substrate surface after laser treatment. The laser beam power was 50% and 90% in the lines 3 and 4, respectively. The magnification increases from 100 times in column 1 to 1000 times in column 3.

Beside the surface treatment, the quality of the adhesive has a significant influence on the bond strength. Taking the linear elastic fracture mechanics (LEFM) into account, small bubbles can affect the crack formation. To influence the density and size of bubbles the adhesive system was mixed on different ways. The bulk samples consisting of the adhesive were examined by light microscopy to characterize the bubbles. The first samples were mixed by hand, the next ones with a mixing machine (speedmixer) and the last ones with the speedmixer after evacuating the pot that contains both components. The manual mixing method produces a high number of bubbles with diameters up to 30µm. No strong improvement has been found using the speedmixer. The density of bubbles has been reduced in the bulk material to 2/3 and the mean bubble sizes has also been decreased to 20µm.

In the samples mixed under vacuum conditions by using the speedmixer, however, no defects bigger than 10μm were found. The curing conditions (room temperature / 5 days; 66°C / 90 minutes) does not affect significant changes in number and size of the detectable bubbles (Figure 3 & 4). After optical analysis lap shear beams were produced and tested under static and cyclic loading conditions. The corresponding results confirm the LEFM suggestions. As a consequence, a higher number and bigger size of bubbles causes lower lap shear strengths and less stress cycles (Figure 5).

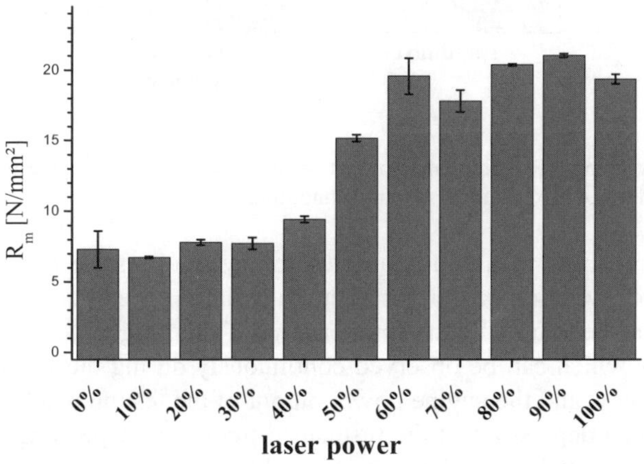

Fig. 2 The amount of the lap shear strength R_m of single overlap samples after laser treatments with different laser beam power.

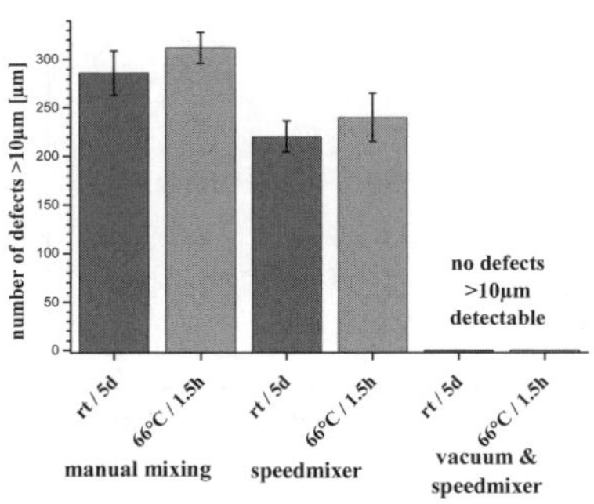

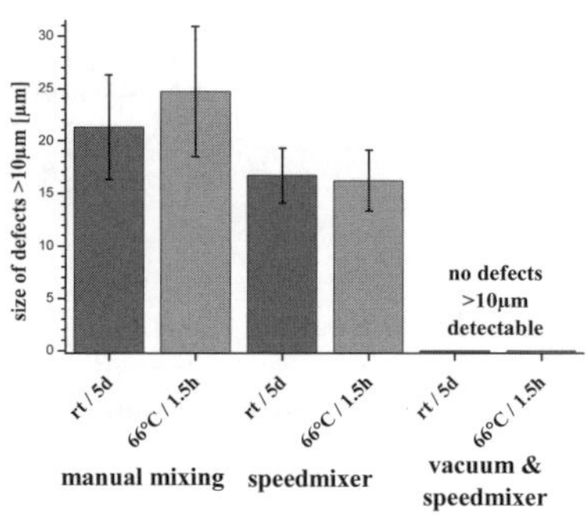

Fig. 3 and 4 Number and size of bubbles in the adhesive bulk sample for different kind of mixing methods (manual mixing, speedmixer, vacuum and speedmixer).

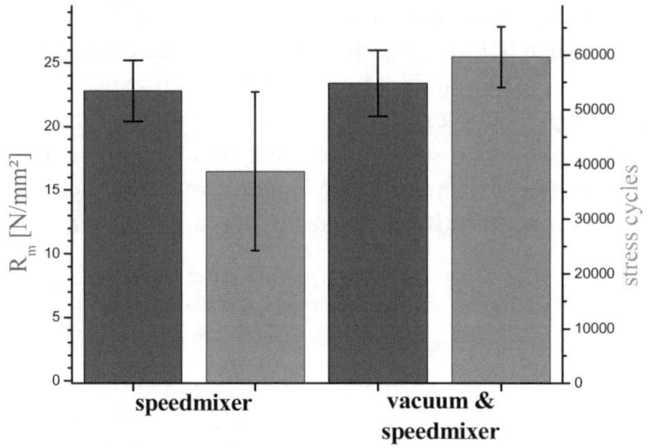

Fig.5 Lap shear strength R_m (blue) and amount of stress cycles (red) for adhesives produced by speed mixing and speed mixing under vacuum conditions.

Starting from these results all samples for cyclic stress tests were produced with a 2K epoxy system mixed under vacuum with the speedmixer. The amount of stress cycles has been used as the final criteria of the fatigue behaviour. As damage parameters which can be observed continuously during the whole test the resonance frequency, and the engine power output of the testing machine were used. During the tests a depression of the resonance frequency depending on decline of the stiffness is detectable. In the same way the engine output ascends when the sample breaks. Consequently, both quantities can be used to qualify the degree of damage prior to the total failure. All of these samples showed a cohesive failure (Figure 6 and 7).

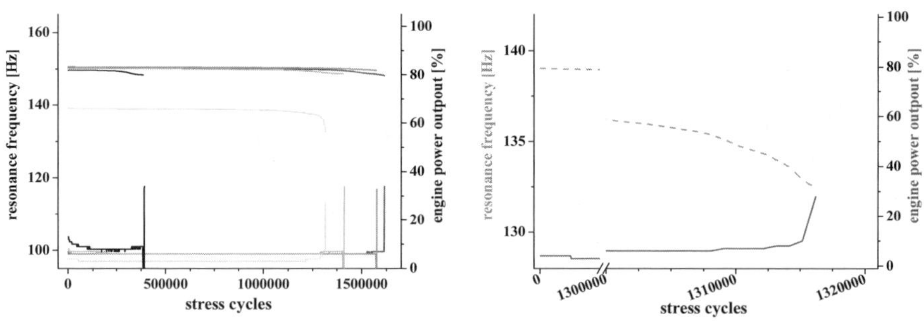

Fig. 6 and 7 Resonance frequency and engine power output vs. stress cycles. The resonance frequency decreases and the engine output ascend.

The failure of samples under static or cyclic loading conditions behaves very different. Under static deformation the adhesive failed in an adhesive or cohesive manner depending on the surface pretreatment. Under cyclic loading conditions not exclusive adhesive or cohesive failure has been detected. Surprisingly under high cycle fatigue the metallic substrate created a fatigue crack near the bond or close to the specimen grips. Consequently, a fatigue crack initiates the failure of the specimen. It was assumed that this type of crack depends of the kind of the grips configuration. In our case bending stresses initiate the fatigue crack at low stress amplitudes. In case of higher stress amplitudes, however, the bonded structure initiate the failure of the specimen as it has been expected.

3. Summary and outlook

Regarding the resonance frequency, the changing of stiffness as well as the engine power output could be important to characterize the fatigue behaviour of adhesive bonded joints. Nevertheless, to go into detail, there is still much more research work to do. The presented damage quantities may help to obtain a secure method for lifetime prediction of adhesive bonded structures deformed under fatigue conditions, even at different environments. Finally, adhesive-bonded structures require an integrated approach, including the design of the construction. It is known that interactions between the used material and adhesives and the surface treatment, influences the stress distribution of the structure. Consequently, the design of the structure must be modified to avoid large stress gradients that are responsible for enhanced damage within the bonded structure. As a first step, it is most important to gain more insight into the above presented mechanisms that are responsible for a safe adhesive bonded structure.

Acknowledgments The authors wish to thank Prof. von Czarnecki, Dipl.-Ing. Holtmannspoetter, Mr. Windsperger, and Mr. Forche for valuable discussions and fruitful cooperation.

References

Brinson H.F. and Brinson L.C. (2008) *Polymer engineering science and viscoelasticity*. Springer, New York

Brockmann W., Geiss P.L., Klingen J. and Schroeder B. (2009) *Adhesive bonding*. Wiley-VCH, Weinheim

Davis J.R. (1993) *Aluminum and aluminum alloys*. ASM International, Cleveland

Habenicht G. (2009) *Kleben Grundlagen, Technologien, Anwendungen*. Springer-VDI, Berlin Heidelberg

ADVANCED MANUFACTURING FOR FIBER REINFORCED METAL MATRIX COMPOSITES (MMC)

M. Silber, M. Wenzelburger, R. Gadow

Universität Stuttgart, Institut für Fertigungstechnologie keramischer Bauteile (IFKB)
Institute for Manufacturing Technologies of Ceramic Components and Composites
Allmandring 7b, Stuttgart, D-70569, Germany
E-mail: martin.silber@ifkb.uni-stuttgart.de
Phone: +49(0)711/685-68-235; Fax: +49(0)711/685-68-299

Abstract The demand for lightweight structures in the automotive industry increases continuously, and the importance of lightweight design principles is also increasing in other industrial branches, aiming towards improved energy efficiency and sustainability. Light metals are promising candidates to realize security relevant lightweight components because of their high specific strength; and amongst them, aluminum alloys are the most interesting materials due to their high plasticity and strain to failure, good processability, passivation in oxygen containing atmosphere, and low cost. However, for many applications, their stiffness as well as strength and fatigue behavior at elevated temperature are insufficient. Metal matrix composite (MMC) formation by integration of reinforcements in the form of continuous fibers can yield a high increase in the alloys specific mechanical properties at room temperature and at elevated temperature. The integration of fibers with conventional manufacturing techniques like squeeze casting, hot pressing or diffusion bonding leads to restrictions in the component's geometry. Moreover, these techniques result in elevated process costs mainly caused by long cycle times and the need of additional protective fiber coatings. In the present paper, an alternative method for the manufacturing of aluminum matrix composites is described, combining thermal spraying and semisolid forming (thixoforging) technologies for the formation of fiber prepregs and subsequent forming with simultaneous densification.

1. MMC production by thermal spraying and thixoforging

The method for manufacturing of aluminum matrix composites described here, combining thermal spraying and semisolid forming (thixoforging) technologies for the formation of fiber prepregs and subsequent forming with simultaneous densification. Therefore, prepregs with the matrix alloy as a thick surface coating on the

J. Wellnitz et al. (eds.), *Sustainable Automotive Technologies 2010*,
DOI 10.1007/978-3-642-10798-6_25,

reinforcement fibers are manufactured in a fast, automated coating process, while reheating, densification and shaping are performed in a separate process, allowing an optimization of both processes towards cycle times and resulting material properties.

For continuous deposition with appropriate tension, the fibers are wound and stretched by using automated devices. The next working step consists of coating the fibers with the matrix material by thermal spraying. The arc wire spraying technique allows the application of a broad variety of aluminum alloys with reduced thermal load on the sensitive fiber substrate. The geometry of the prepregs has to be adapted to the design of the final component. Then, prepregs are trimmed and laminated prior to rapid heating by infrared or inductive devices. The prepregs are finally processed to the MMC component by thixoforging (see Figure 1).

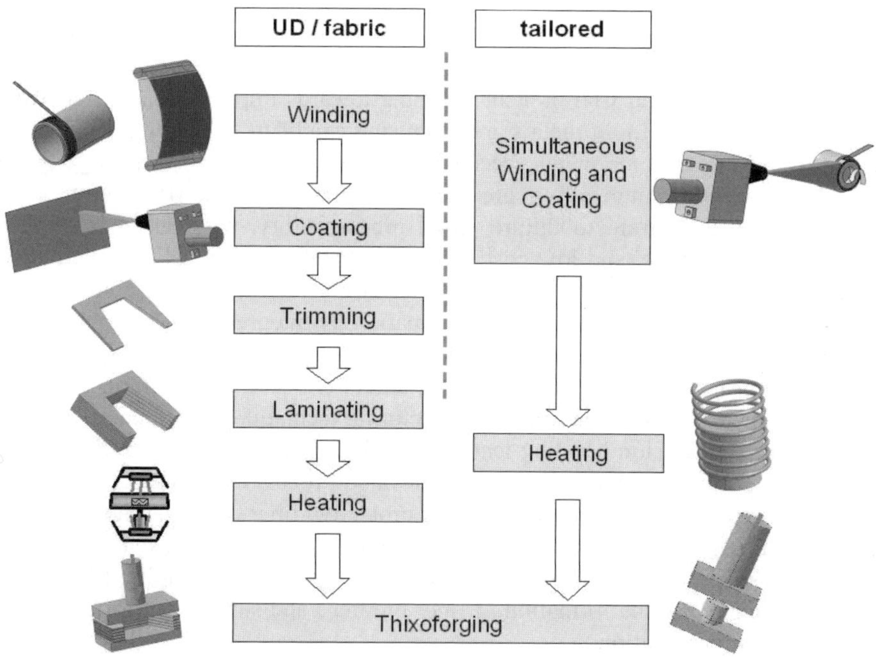

Fig. 1 Process chain for the manufacturing of fiber reinforced MMC components by thermal spraying and thixoforging.

1.1 Fiber coating by thermal spraying

For deposition of the light metal alloy, the industrially well-established twin wire electric arc spraying process is applied. With its high deposition rate and low heat transfer to the substrate, this technique is particularly suitable for the deposition of matrix material on fiber layers for prepreg production.

A constant electric arc is generated between two continuously fed wires consisting of the coating material. The melt, which has temperatures of up to 2000-3000° C, is atomized and sprayed towards the substrate by compressed air (see Figure 2). The deposited material is quenched and solidified on the substrate surface within 0.1-10 μs (Pawlowski 1995). The continuous character of the thermal spray process and the small droplet size, in combination with optimized speed profiles for handling devices and advanced cooling systems, lead to low thermal load of the sensitive fiber substrate. Therefore, the substrate material is not damaged even for high melting point spray materials (Gadow and von Niessen 2003). This technique enables the production of composites with different matrix/fiber volume ratio by varying the coating thickness.

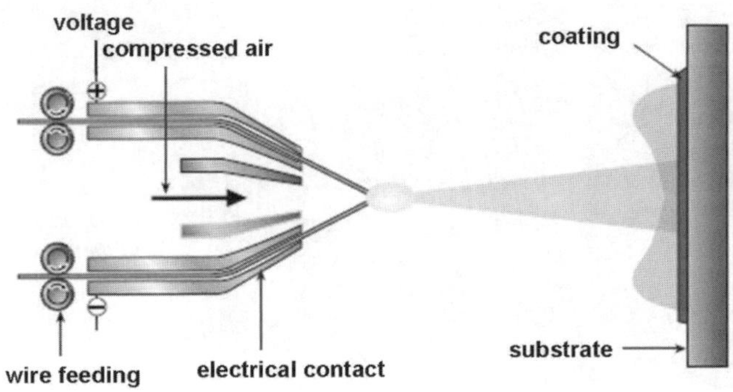

Fig. 2 Principle of the wire arc spraying process (Sulzer 2005).

1.2 Heating and thixoforging

The piled coated prepregs are heated to the semisolid temperature for thixoforging, i. e. 577-590 °C for Al 6Si, providing a liquid fraction of 40-60 vol% (Uggowitzer *et al.* 2000). To reach the required temperature window for semisolid processing a short wave infrared heating device with high-power IR radiators was applied for reheating of the fiber prepreg materials. For the thixoforging process, a hydraulic, uniaxial high-speed press was used with a maximum forging and locking pressure of 5,000 kN and a maximum punch speed of 800 mm/s. The high punch speed guarantees appropriate thixotropic behavior of the matrix material, which rapidly cools during transfer from the heating device to the press, but also in the die as a result of heat transfer from the prepregs. The die was heated to a temperature of 450 °C in order to limit early cooling and solidification of the material, which leads to an increase in the viscosity of the material and shear stresses, and finally, to damage or disarrangement of the fibers. The complete forging stroke with an adapted speed profile lasted 20 s (Siegert *et al.* 2004). Figure 3 shows the microstructure of a composite consisting of 18 layers of carbon fibers coated with Al Si6 by arc wire spraying. The good wetting and complete infiltration of the metal matrix material into the fibers was experimentally determined. The single fibers show no deformation or other mechanical damage. Experimental evaluation also showed only minor chemical reaction (and thus damage) at the fiber/matrix interface.

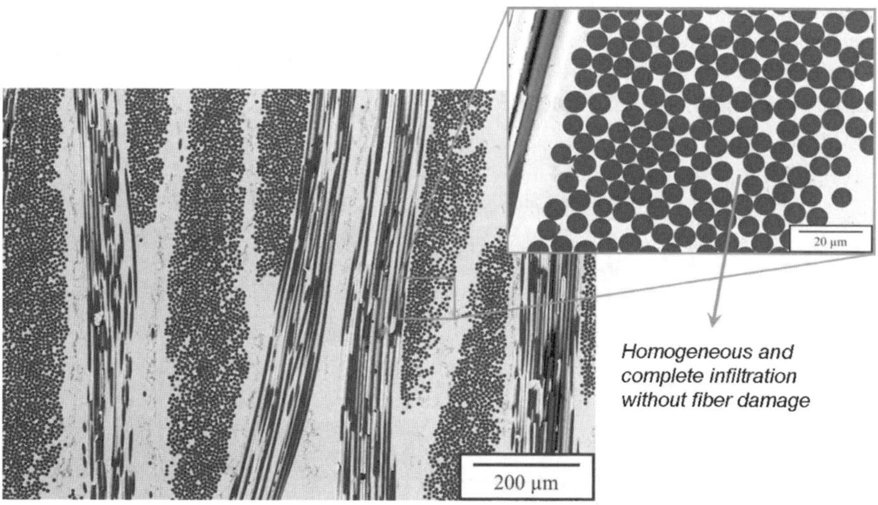

Fig. 3 Microsturcture of Al Si6 MMC with woven carbon fiber reinforcements.

2. Equipment for producing advanced fiber reinforced MMC

2.1 Production of unidirectional fiber reinforced composites

For manufacturing of unidirectional fiber reinforced light metals, winding of the continuous fibers to fiber bundles is required. Regarding technical equipment, a winding unit was developed with the focus on short cycle times during prepreg manufacturing (see Figure 4). By means of an adapter, different commercial fiber coils can be mounted. A friction brake unit at the rotational coil axis keeps the fiber under defined tension during the winding process. The fiber guide unit can be also continuously varied in speed and horizontal range of the winding width. The electric drive system for the fiber guide defines the fiber strand position on the cylinder in horizontal direction.

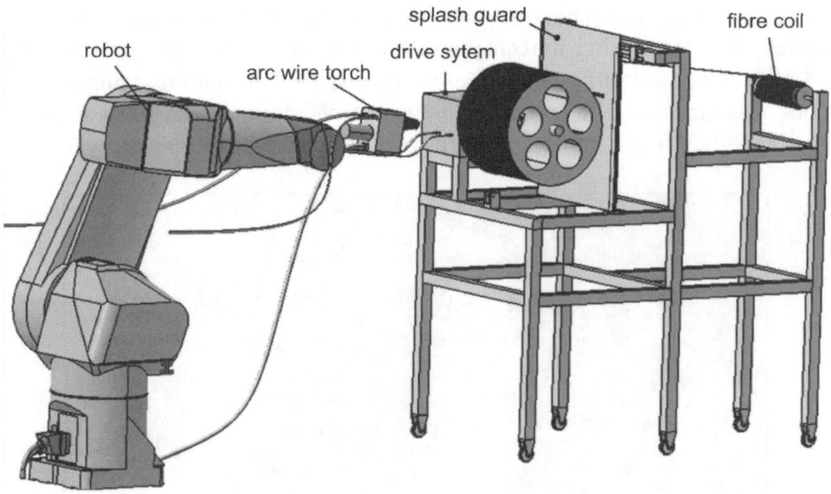

Fig. 4 Robot guided spraying torch and winding unit.

For deposition and winding, cylinders of different size can be used. By adaptation of the cylinder diameter to the coating thickness, it is possible to produce flat fiber prepregs. After removing the coated prepregs from the cylinder and cutting them to the desired fiber length, the primary cylindrical shape deforms towards a flat geometry due to the residual stresses which arise during coating deposition. A flat fiber prepreg structure reduces the handling effort in the subsequent processing steps.

The relative speed between wound fibers and particle jet during the coating process depends on the rotational speed of the cylinder. Thus, the axial feeding rate of the spray torch has to be adapted to the cylinder's rotational speed. In order to keep the thermal and kinetic load on the fibers as low as possible and to minimize cycle times, the relative speed has to be optimized (Gadow *et al.* 2004).

2.2 Production of 2D textile fiber reinforced composites

Two-dimensional fiber structures are required for components subjected to complex mechanical load in large area applications. In this case, textile woven fabrics are used as basic material for coated prepregs. Special winding equipment is needed to fix and stretch the fabric in a way that enables continuously coating from “coil to coil”. Load cells inside the drive rolls of the winding equipment control the tension of the stretched fabric. If elongation or contraction of the fabric due to thermal expansion occurs, electric drives will compensate the change of tension within a control loop. With this equipment fabric rolls up to a web width of 1,500 mm can be coated continuously with a predefined pre-stress in one single procedure. The meandering movement and kinematics of the spray torch is performed by a 6-axis robot system. Figure 5 illustrates the winding equipment and the robot guided torch. The coated prepregs can be trimmed in a subsequent step by material NC-cutting control systems providing reduced material wastage.

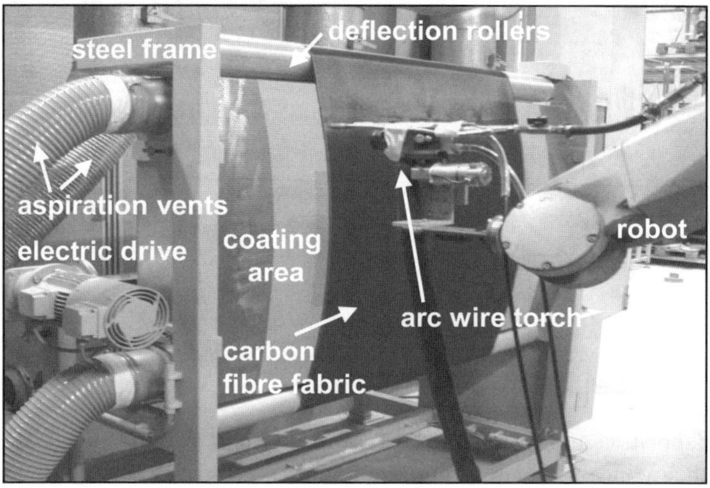

Fig. 5 Twin wire electric arc coating process and winding equipment for continuously coating of textile fiber fabrics.

2.3 Production of tailored, local reinforced MMC

For local improvement of components mechanical properties, the manufacturing of a wheel rim with local reinforcement in the central area, see Fig. 6, is presented as a case study, for which a suitable processing technology for semi finished material was developed. Thereby, the aim was to produce semi finished parts with continuous fiber reinforcement around the hub, with the fibers being placed on their final intended position in the component. Therefore, the fibers should be well embedded in the matrix material, with the unreinforced material of the billet flowing around the reinforcement without disarranging it geometrically during thixoforging.

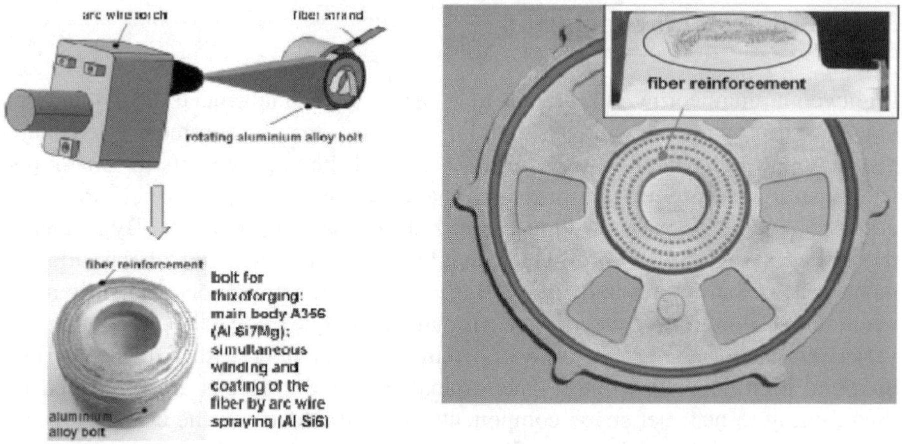

Fig. 6 Principle of simultaneous winding and coating, left; fiber reinforced rim manufactured by thixoforging; cross section of the centre with carbon fiber reinforcement, right.

The local reinforcement consists of HT carbon fibers. During winding of the fiber on semi finished billets (A356), the fiber strand was coated with Al Si6 alloy. The locally reinforced, semi finished billet was heated by inductive heating prior thixoforging. For continuous winding and coating a step in the billet is used for application of the reinforcement, leading to a finally cylindrical semi finished part with a pre-impregnated, fiber reinforcement integrated in the billet top, see Fig. 6, left. The unreinforced material of the billet delivers the metal material for complete mold filling of the rim die. During processing of semi finished billet, there are three challenges to overcome:

- the fiber strand has to be spread exactly to the width of the desired reinforcement area (step in the billet), choosing an appropriate fiber strand thickness in order to obtain a thin, well impregnating fiber layer,
- the winding process has to be applied to the desired tension of the fibers in the matrix, thus being able to take influence on the final state of residual stress in the composite area, and
- the coating process itself requires adaptation of the coating area, coating distance and feed rate in order to obtain the pre-defined fiber volume content; this leads to the requirement of further adaptation of the cooling system to control deposition temperature.

After processing of the semi finished parts, the cylinders were heated up by an inductive heating device and finally thixoforged to the near net shape component (wheel rim). During the thixoforging process, the fiber reinforcement was disarranged to a certain extent, see Fig. 6, right, detail. However, work on this subject is still in progress and requires further research and optimization with the aim on better control of form filling and flow paths during the densification process.

3. Summary and conclusion

Lightweight engineering is a crucial field of research in advanced design and engineering of components and systems for high performance automobile technology. Although light metal MMC can fulfil most of the requirements in system development, new production processes have to be developed to ensure cost-effective manufacturing as well as optimized mechanical properties. By means of thermal spraying and thixoforging technologies, MMC with UD fiber reinforcement, 2D fiber structures and tailored local reinforcements can be efficiently manufactured. The twin wire electric arc spraying technique allows the deposition of a variety of casting aluminum alloys on reinforcement fiber with reduced thermal load and high deposition rates. The laminated coated prepregs are processed by thixoforging to near net shape components. With the focus on the critical process parameters (flexible production, fiber cooling and tension during coating), advanced manufacturing systems for each reinforcement strategy were developed. Thus, composite materials which fulfil the requirements of high performance applications can be made available.

Acknowledgments The authors thank the Ministry of Science, Research and Art of the State of Baden-Württemberg and the Stiftung Industrieforschung e. V., Cologne, for financial support of this work. Furthermore, the authors thank the Institute for Metal Forming Technology, Universität Stuttgart, for thixoforging experiments.

References

Gadow R. and von Niessen K. (2003) Ceramic Transaction 151, 3-18

Gadow R., von Niessen K. and Unseld P.(2004), "Thixoforging of continuous fiber reinforced AlSi/AlMg-alloys", edited by P. Bernasovsky and P. Brziak, Int. Conf. Advanced Metallic Materials and their Joining, Bratislava, Slovakia

Pawlowski L. (1995) The science and engineering of thermal spray coatings, Chichester: J. Wiley

Siegert K., Gadow R., von Niessen K., Unseld P. and Speicher M. (2004), in: Proc. 8th Int. Conf. Semisolid Processing of Alloys and Composites (S2P 2004), September 21-23, Limassol, Cyprus (Pub. NADCA Wheeling, Illinois, USA, 2004), Paper # 23-3

Sulzer; http://www.sulzermetco.com (2005)

Uggowitzer P., Gullo G.-C. and Wahlen A. (2000) in: Proc. Int. Conf. \`Vom Werkstoff zum Bauteil´, edited by H. Kaufmann, P. Uggowitzer (LKR-Verlag, Randshofen), p. 95-107

OPPORTUNITIES AND LIMITS FOR COMPOSITES IN CARS

C. Ruegg

Chris Ruegg Innovation Management,Zurich, Switzerland
E-mail: chris.ruegg@chris-ruegg.ch

Abstract In a view of environmental considerations the question is: Will carbon fiber composites be the material for volume car production in the near future? Unfortunately, despite of the great weight savings potential, high material cost is still a major barrier for carbon fiber composites in mass production. The use of light weight carbon fiber composites in future cars will depend on low cost carbon fiber technology combined with innovative new car design concepts adapted to cost efficient manufacturing processes.

1. Introduction

Over the years we have seen an impressive number of high performance sports cars made with carbon fiber reinforced plastic materials (CFRP). CFRP composites are very light in weight, stiff and crash resistant, Figure 1.

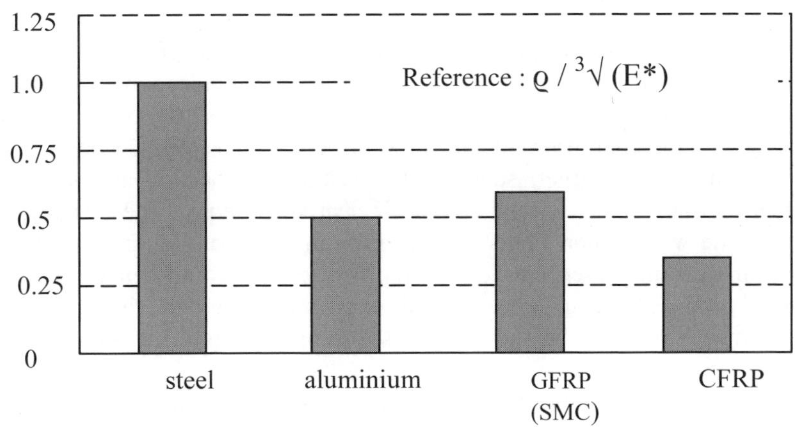

Fig. 1 Mass comparison of various materials for same bending stiffness.
(E*= modulus of elasticity for quasi-isotropic arrangement of fibers)

J. Wellnitz et al. (eds.), *Sustainable Automotive Technologies 2010*,
DOI 10.1007/978-3-642-10798-6_26, © Springer-Verlag Berlin Heidelberg 2010

In a view of environmental considerations the question is: Will carbon fiber composites be the material for volume car production in the near future? Unfortuna-tely, despite of the great weight savings potential, high material cost is still a major barrier for the use of carbon fiber composites in mass production, Figures 2 and 3

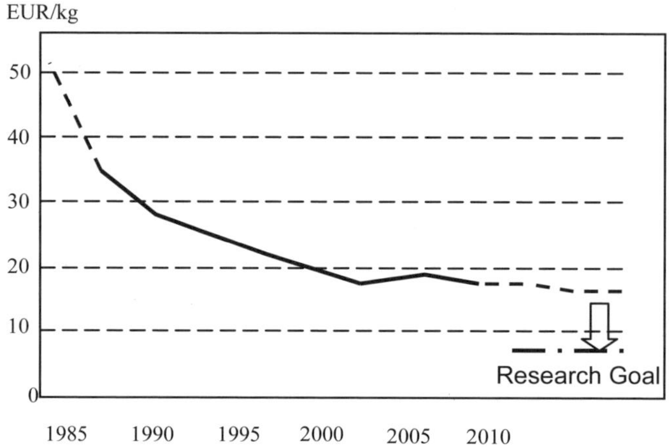

Fig. 2 Historical cost development of carbon fibers and research goal (5$/lb).

2. Cost development of carbon fibers and transformation processes

Over the years carbon fiber producers have made enormous efforts to reduce the cost of carbon fibers and they have been quite successful indeed, but material cost is still high compared to steel or aluminium, Figure 2. There are research projects underway with the goal of producing carbon fibers at a cost of 3-5$ per pound. In order to achieve this target, substantial investment is needed in research and equipment for new precurser and carbon fiber production processes. Today, components made with carbon fibers are generally more expensive than components made with steel and, even considering the lower density and hence lower mass needed to produce an equivalent strength chassis or body part, the material cost alone, in most cases will be above the cost of components made with steel, Figure 3.

Great efforts have also been made to develop low cost transformation processes for carbon fiber components viable for large volumes. But, even with very efficient processing methods carbon fiber composite components in most cases will be substantially higher in cost than their steel or aluminium counterparts, unless such components incorporate other new benefits in addition to pure weight savings.

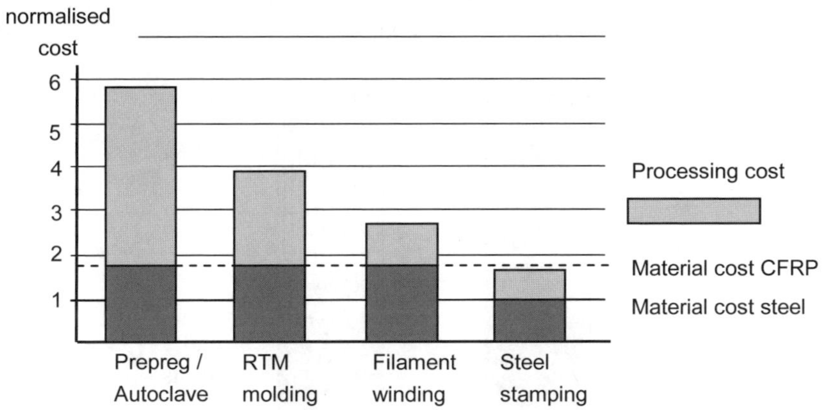

Fig. 3 Cost comparison of components made with carbon fibers vs steel.

3. Special case : carbon fiber propeller shafts

Carbon fiber propeller shafts are one example where, due to the high specific modulus of elasticity of carbon fibers, a typical two-piece steel shaft can be substituted by a one-piece composite shaft performing at the same maximum speed of rotation. As a result, such carbon fiber propeller shafts have the potential to achieve substantial weight savings over steel shafts at equvalent total cost, Figures 4 and 5.

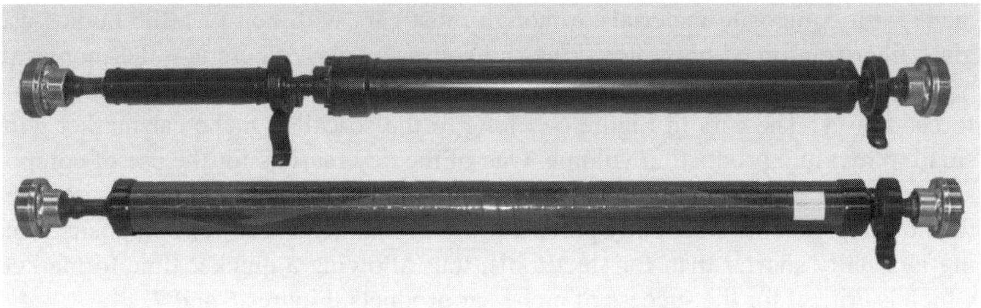

Fig. 4 One-piece CFRP propeller shaft vs two-piece steel prop shaft.
(courtesy xperion GmbH)

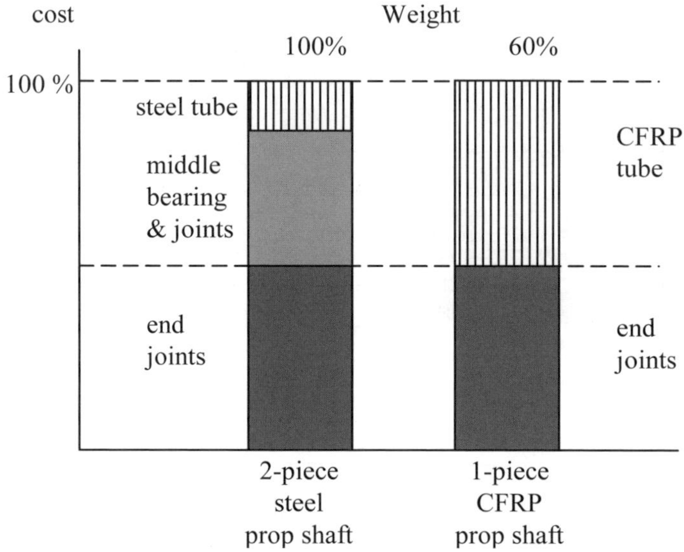

Fig. 5 Typical weight and cost comparison for a one-piece CFRP prop shaft vs two-piece steel prop shaft.

4. Tooling cost versus material and processing cost

Despite of the cost barrier mentioned above, there is an important number of cars made with composite materials, although most cars with non-metallic bodies use glass fiber reinforced polyester skins and some use glass fibers in a thermoplastic matrix. This class of material is substantially lower in cost than carbon fiber reinforced epoxy. The cars in Figure 6 belong to the so-called niche car market with small to medium production volume. One of the key reasons for the use of composites in niche cars is tooling cost. Tooling cost for composite body parts is lower than tooling cost for steel body parts. On top of that, lead time for composite tooling is usually shorter than for steel tools, thus allowing a quicker time to market, which is critical for the success of niche car products, Figures 6 and 7.

Fig. 6 Selection of cars with composite body components.

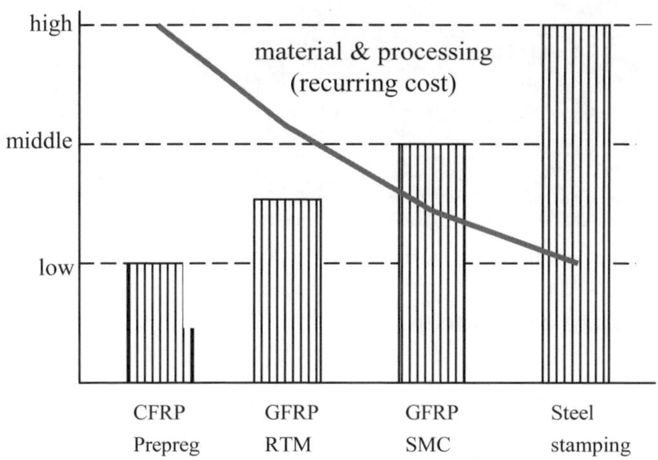

Fig. 7 Tooling cost vs material cost.

The preference and logic for material selection was demonstrated in the history of the Renault Espace from its early years till today. The first model, the Espace I, was launched in 1984. The Espace was one of the first minivans in the market and there was much uncertainty about the number of customers that would buy such a car. Hence, Renault's planning was aimed at moderat quantities, at least during the initial years. The Espace I had a car body made with glass fiber reinforced polyester skins manufactured by the so-called resin transfer molding (RTM) method, where tooling cost is low. Soon, sales of the Espace proofed quite successful and with increased volume production was switched from the RTM process to the so-called sheet molding (SMC) process. Tooling cost for sheet molding is higher than for resin transfer molding but material cost is lower. Then the Espace became even more successful and today's model has again a steel body, Figure 8. The Renault Espace is a good example of the importance of material versus tooling cost in relation to volume. There are many examples of cars with RTM or SMC components during market introduction in order to avoid the risk of high tooling investment until sales figures justify steel tooling. Needless to say that the Espace is a much heavier car today than its earlier composite models

Fig. 8 Renault Espace, models I, II, III and IV (from 1984 to 2008).

5. Trends and future car concepts

Looking to the future of materials in automotive engineering we may see the following trends:

Economics in production call for as much standardisation as possible, while customers want more individualisation. Hence, a successful car concept uses standardised components for non-visible parts like chassis components, whereas body parts should be easily adaptable to meet changing styles and customer preferences for individualisation. Environmental responsibility calls for smaller cars with electric or Hydrogene energy storage and recyclable materials. We may also see a shift from the very high passive safety features during the 90's towards increased active safety features in the future.

Table 1 Trends for future car concepts.

	trends and needs
Economics /Production	standardisation of non-visible components, e.g. chassis
Marketing & Sales	individualisation of vissible parts, e.g. body and interior
Environment	1. low or zero emission Engine efficiency Lower vehicle mass - smaller cars - light weight materials electric / hybrid / H_2 / fuel cells / . . 2. Recyclable materials
Safety	shift from high passive safety to increased active safety

In order to make efficient use of composite materials in automotive engineering, it will be necessary to make some changes with respect to design concepts. Current car structures are engineered around process technologies for steel. Future composite car structures will have to be adapted to allow the use of components made with cost effective manufacturing processes for composites, particularly that of carbon fiber composites.

The concept study in Figure 9 is one example of a car structure incorporating a composite box beam floor as a new chassis component in a car. Such a box beam floor would combine high torsional stiffness with high crash resistence and serve as a container for battery racks and hydrogene storage tanks, etc. It would be produced by the so called pultrusion process, which is a well established low cost manufacturing process for composite structures.

It is clear that the concept presented in Figure 9 is but one of many ways in an effort to incorporate composite materials in cars. However, it demonstrates that car designers are looking for innovative new ways to take advantage of the high weight saving potential of carbon fiber composite materials in cars of the future.

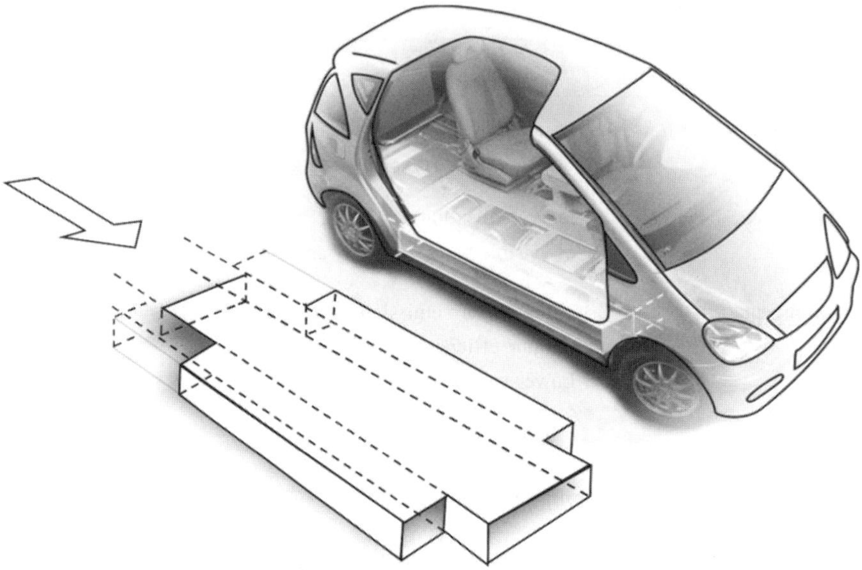

Fig. 9 Chassis concept with pultruded composite box beam floor. Concept drawing ITD.

6. Conclusion

As a conclusion we can state that the use of light weight composite materials, particularly that of carbon fiber composites in future cars, will depend on low cost carbon fiber technology combined with innovative new car design concepts adapted to cost efficient manufacturing processes.

Part 6
New power train technologies

FORMULA HYDROGEN – INTERNATIONAL DIMENSIONS OF THE HYDROGEN POWERED RACING CAR PROJECT

A. Subic[1]**, J. Wellnitz**[2]

[1]RMIT University, Bundoora, Victoria 3083, Australia
Email: aleksandar.subic@rmit.edu.au

[2]University of Applied Sciences Ingolstadt, Germany
Email: joerg.wellnitz@fh-ingolstadt.de

Abstract RMIT University and University of Applied Sciences Ingolstadt have unveiled together the Australia's first hydrogen-powered racing car end of 2008 in Melbourne, in a pioneering project that demonstrates the possibilities of hydrogen as the clean, renewable fuel of the future. The car will be bidding for the title of world's fastest hydrogen-powered racing car when it attempts to break the Guinness World Records mark for speed by a vehicle of its class in 2010. This research and development project has commenced in 2006 through collaboration between the two universities with the two research teams working concurrently at different parts of the world. The paper presents the main features of this innovative international project and the outcomes achieved.

1. Introduction

The Formula Hydrogen project represents an international collaborative research and development project involving RMIT University in Melbourne and University of Applied Sciences in Ingolstadt. The project commenced in 2007 with the two teams working 24 hrs per day at different sides of the world (e.g. when Australian team went to bed, German team started work). The teams came together mid 2008 in Australia for the purpose of vehicle system integration, trials and demonstration.

Researchers from both institutions initially developed the technology road map for the project and in March 2007, the hard work of building the racing car of the future began. Following the project team model of the Formula SAE, Mechanical Engineering students were put in charge of the key hands-on roles, supervised by staff from the two universities. While RMIT focused on constructing the chassis and body as well as vehicle system integration, the German group worked on the redevelopment of the standard, two-cylinder Rotax motorcycle engine to run on hydrogen gas (see Figure 1). As hydrogen burns less effectively than conventional

J. Wellnitz et al. (eds.), *Sustainable Automotive Technologies 2010*,
DOI 10.1007/978-3-642-10798-6_27, © Springer-Verlag Berlin Heidelberg 2010

fuel, the remodelled and retrofitted engine had to take into account the loss of torque and performance. A supercharger and a sub-system that uses liquid nitrogen to cool the fuel were integrated into the design.

Storage of hydrogen gas in the vehicle is of key concern. Hydrogen atoms are very small particles, which are highly permeable, and require a complex storage solution. We used in this vehicle a seamless thin walled aluminium pressure tank wound with carbon fiber to reach high density and stability at lower weight (see Figure 2). This tank has to resist very high pressure, which is needed for filling. For example, at least 350bar of pressure is required to carry 3kg of hydrogen fuel. If we consider that one kilogram of fuel lasts for around 100 km, this means that we would need a pressure of about 700bar to reach an acceptable range compared with conventional vehicles, which is what makes this storage problem a complex one. Fortunately, for a racing vehicle such a range is not required hence we were able to compromise by implementing a storage tank at a lower pressure of 200bar, which allows a filling of 72 liters or 1,075 kg of hydrogen respectively.

In order to reduce the overall weight of the car, the entire body was developed and manufactured from carbon fibre composites at the RMIT Composites Laboratory in Melbourne (see Figure 3).

This paper will describe in detail the research and development behind the Formula Hydrogen car and the outcomes achieved by the research teams from the two universities.

2. Formula Hydrogen Design

For us, the choice of hydrogen as the principal energy carrier was strategic. Hydrogen is the most common element in our universe that can be produced from water, natural gas and biomass using renewable energy sources; in relation to weight the energy density of hydrogen is nearly three times that of petrol; and it is virtually emission-free, whereby the exhaust emits only water vapour.

Like in a standard fuel-injected engine, in Formula Hydrogen the fuel is fed into the system via a set of twin injectors. Unlike a standard combustion engine, the fuel stored in the tank is highly compressed hydrogen gas, stored at 200 times atmospheric pressure. When released from the tank, the hydrogen runs through a series of pressure-reducers and check valves so by the time it reaches the engine, the gas has been brought down to about five times atmospheric pressure. The modified combustion engine burns the hydrogen to produce the energy needed to power the racer but unlike petrol, diesel or natural gas-fuelled cars, the only by-product pumped out of the twin exhaust pipes is harmless water vapour.

Safety is a key concern with any fuel, but particularly when dealing with highly-flammable hydrogen gas. Complex connection technology was developed to ensure the safety of drivers and the crew, including specially-designed elements such as hydrogen detectors, sequential fuel injectors, connectors and adaptors. These systems measure pressure and temperature in the exhaust pipe, at the cylinder head and in the tank – before and after fuel injection. With so many parameters to calculate, a fast and complex engine management system is required to keep track of the huge data flow and the car uses the MoTeC system that is one of the largest that can fit in a racer of its class.

The storage of the alternative fuel also required a complex solution because hydrogen is highly permeable at atomic level. The seamless, thin-walled aluminium tank was wound with carbon fibre to increase density and stability, while keeping down the weight. The tank needed to be strong enough to resist the extreme pressures involved when re-fuelling the car. It can hold up to 1,075 kg of compressed hydrogen, or 72 litres, giving the racer a range of about 100km.

Engine:
Manufacturer: Rotax
2-cylinder in-line, capacity 798 cm³
Weight: 63 kg
Power output:
Petrol: 62 kW/ 85 HP @ 8.000 min-1
H2: estimated around 40 kW*;
2 Transmission setups (acceleration and maximum speed)
*increased power with implementation of supercharging and intercooling

Fig. 1 Modified internal combustion engine used in Formula Hydrogen.

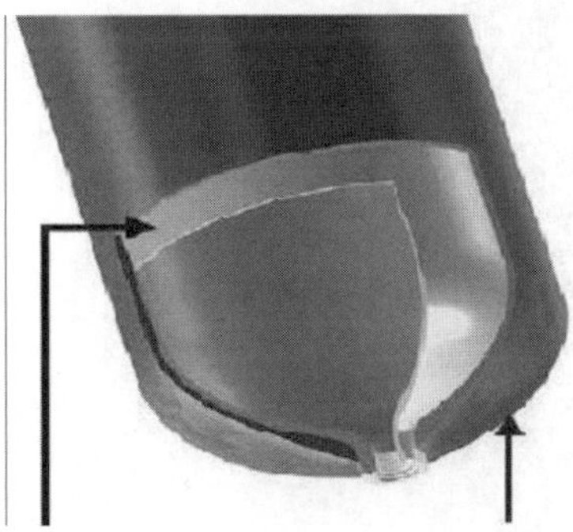

Hydrogen Fuel System:
Tank manufacturer: Dynetek Europe GmbH
High Tech Carbon Fibre/Aluminium Tank
Nominal weight of H2 tank: 28 kg
Pressure: 200 bar (gaseous H2)
Hydrogen capacity: 1 kg (72 liters) of H2
H2 safety system incl. Hydrogen sensor, pressure regulation and control devices
Hydrogen fuel injectors: ported fuel injec-

Fig. 2 High pressure storage tank used in Formula Hydrogen.

Fig. 3 Formula Hydrogen developed by RMIT University and University of Applied Sciences Ingolstadt.

Fig. 4 Formula Hydrogen on show in Melbourne.

3. Outcomes

The Formula Hydrogen project has achieved a number of important outcomes as described below.

a) Two universities established a strong collaboration in sustainable automotive engineering leading to the development of a novel hydrogen powered racing car, the first of its kind in Australia, aiming to be the fastest in its class.
b) Essential part of this project was broad community engagement both in Australia and in Europe to showcase the car and to promote green car technologies and future projects (see Figure 4: Formula Hydrogen on show in Melbourne, Australia). The car was recently showcased at the Audi Innovative Industry Expo in Ingolstadt and is invited to Melbourne Formula One Grand Prix in 2010.
c) The project demonstrated new possibilities of modified internal combustion engine technologies in conjunction with hydrogen fuel.

The two universities aim to commence work on Mk2 of the Formula Hydrogen by potentially further redeveloping the internal combustion engine for multi-fuel applications (e.g. hydrogen with natural gas; biofuels with natural gas or hydrogen)

ANALYSIS OF POTENTIAL INCREASES IN ENERGY EFFICIENCY FOR PISTON COMBUSTION MACHINES WITH UNCONVENTIONAL GEOMETRY

A. Gotter[1], B. Schapiro[2]

[1]gofficient, An der Schmackertz Kull 9, D-41564 Kaarst, Germany
E-mail: info@gofficient.de

[2]RKM, RotationsKolbenMaschinen GbR, Schloss Str.30, D-12163 Berlin, Germany
E-mail: boris@rkm-schapiro.org

Abstract Starting with various alternative piston combustion machine (pcm) designs, such as RKM Schapiro piston machines, potential increases in thermodynamic potential based on changed boundary conditions are estimated. The boundary conditions of conventional pcms define the parameters maximum peak pressure, compression ratio and friction mean effective pressure. Adding a supercharger to a modern conventional pcm determines the charge pressure ratio parameter. This study estimates changes in thermodynamic potential by varying the boundary conditions affecting machine parameters beyond those defined by conventional pcms. This was accomplished using SimEngine, a software application for simulating engine thermodynamics. This software has great functionality for simulating unconventional combustion engines and engines equipped with additional thermodynamic components.

1. Scope of Analysis

Modern internal combustion engines have a total of more than 100 years development and optimisation time passed and can surely described as well product – but are they really optimal?

1.1 RKM Schapiro engine

There are some innovative alternative designs, such as the RKM Schapiro engine, which may have potential to beat the weaknesses of conventional engine designs. Taking the RKM engine as example, the limitations of maximum peak pressure and sealing can be completely solved. However, due to high peak acceleration forces, the maximum engine speed is limited.

J. Wellnitz et al. (eds.), *Sustainable Automotive Technologies 2010*,
DOI 10.1007/978-3-642-10798-6_28, © Springer-Verlag Berlin Heidelberg 2010

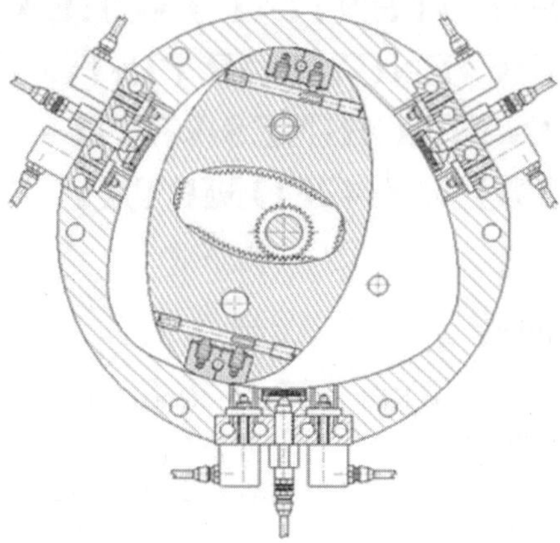

Fig. 1 Principle schematic of one of possible 1-axis RKM engines.

1.2 Targets

The target of this article is to identify thermodynamic potentials of exceeding current boundary conditions. So, this article should not qualify the RKM concept itself. It shall give an overview of the potential which can currently not be used.

The parameters, limited by current technology, are maximum peak pressure (pmax); compression ratio (CR) and friction mean effective pressure (fmep). Table 1 shows an overview of the varied parameters.

Table 1 Overview of varied parameters.

Parameter	Current limitation	Investigation of up to…
Maximum peak pressure	~120 bar (Otto)	400 bar
	~200 bar (diesel)	
Compression ratio	~20 (diesel)	50
Break mean effective pressure	~25 bar	100 bar (a.m.a.p.)
Friction mean effective pressure	~0.5 bar	0.1 bar

A second part gives an overview of applications for alternative compressors or expanders. An investigation subject may be a steam expander application.

1.3 Analysis of wasted power

Based on the principle laws of thermodynamic, the internal efficiency (η_i) of an internal combustion engine is limited by its compression ratio and the isentropic coefficient of the working fluid (air):

$$\eta = 1 - \varepsilon^{1-\kappa},$$

with η efficiency, ε compression, and κ isentropic coefficient, $\kappa > 1$.

It can be seen that an improvement of the compression ratio directly leads to a higher efficiency. The isentropic coefficient of combustion gases is dependent from λ, the relative air/fuel ratio, which is much better for lean mixtures than for rich mixtures.

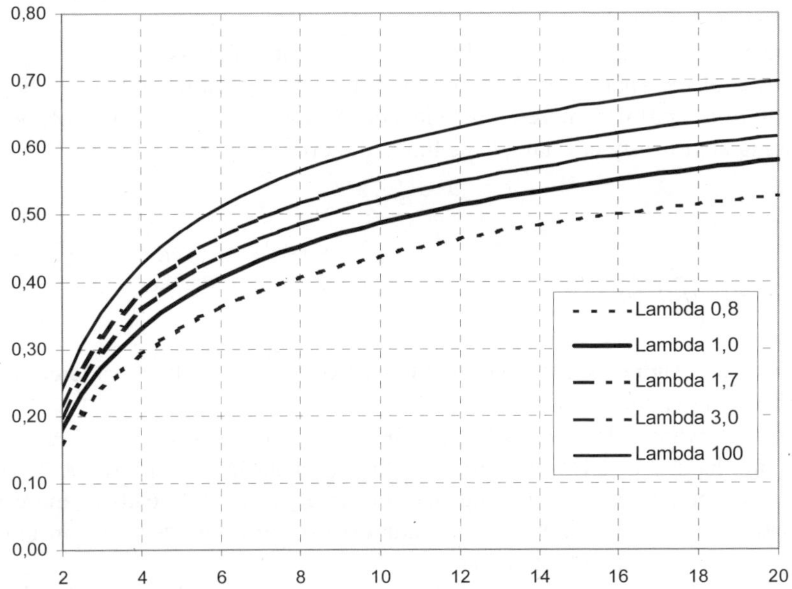

Fig. 2 Thermodynamic efficiency potential depending on compression ratio and = A/F ratio.

However, the compression ratio of Otto engines is limited by knocking behaviour of the gasoline fuels. It is also not possible to operate an Otto engine very lean, so its basic thermodynamic potential seems to be exhausted.

Diesel engines already have a high compression ratio of ~ 20 and also operate with lean mixture. This is the reason for there higher efficiency. However, only a few diesel engines have higher compression ratios due to the limiting increasing maximum pressure and increasing friction due to heavier engine design. The current resulting effective efficiency (η_e) is in the range of 40%.

1.4 Gasoline applications and Miller principle

For gasoline applications, there is one principle available to avoid knocking while increasing the compression ratio: the Miller principle. Hereby the valve timing is trimmed in a way that the compression is lower than the expansion ratio. A geometrical compression ratio of 13-15 can be reached without knocking, but it is necessary to boost the engine to achieve an acceptable specific power. This principle is used in some modern block heat and power plants and raises their efficiency a few percent.

However, knocking is a phenomenon which is caused by high temperatures and pressures. The high temperatures can be avoided by decreasing the geometrical compression ratio.

If the engine power output should stay constant, it is necessary to boost the engine and cool the charge air down to an ambient temperature like level. So, the temperature at end of compression (relevant for knocking) can kept cool, but the pressure is kept nearly constant. This limits the efficiency gain by the Miller principle.

1.5 Diesel applications

Modern diesel engines are limited mainly by two factors:the fuel evaporation and the maximum cylinder pressure.

Poor fuel evaporation causes big droplets, which leads to high soot emissions on the one hand and NOx emissions on the other hand, caused by locally high combustion temperatures. Avoiding that unwanted emissions leads to more lean combustion, which costs power, this trade-off can be optimized with increasing fuel pressure levels of up to 2000 bar.

The other limiting factor, the maximum cylinder pressure is given by the engine geometry. If a high engine power output is wanted, either the compression ratio has to be decreased or late injection timing has to be applied, both costs efficiency.

By that reason, the diesel engine was identified as the engine with the highest thermodynamic potential, given by new engine geometries.

2. Simulation results

All following simulations were calculated with SimEngine, the thermodynamic simulation software with advanced thermodynamic functionality for internal combustion engines. As one example, the software can handle numerous of real gas media, a behaviour which is important especially at high pressures. Furthermore, that software can also handle steam processes with different media in a combined simulation environment with internal combustion engines.

2.1 Base parameters

Starting from a well optimised series engine, several modifications have been calculated. Most of these modifications are not applicable at an engine with standard geometry, either due to the compression ratio or due to the maximum cylinder pressure.

The base engine has an effective efficiency (η_e) of approximately 42.2%, which is an actual best reference case for passenger car diesel engines. All simulations are based on 1500 rpm engine speed and 95°C water and oil temperature.

2.2 Injection timing

The first variation loop is the injection timing. Starting from the base case, the injection timing was modified in direction early. Observed was the indicated an effective efficiency as well as the maximum peak pressure.

As first result, it can be seen, that the maximum peak pressure increases dramatically with the injection timing, while the indicated mean effective pressure (imep) stays nearly constant.

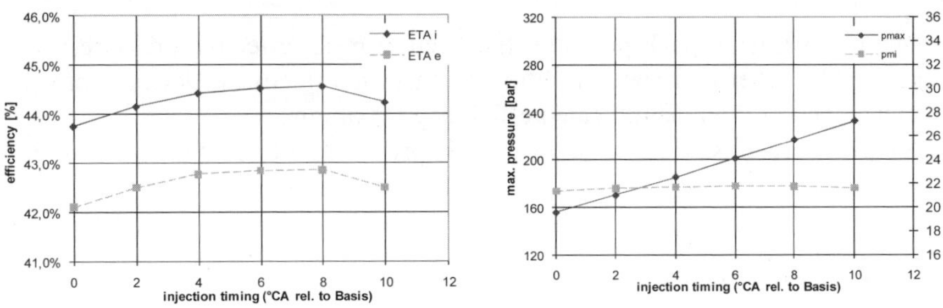

Fig. 3 Effect of injection timing.

As moving the injection timing in early position, efficiency first raises, but at too early injection timings, the efficiency drops due to increasing wall heat transfer. A reasonable value for the injection timing is 4° crank angle (CA) earlier than the base scenario, which leads to ~0.5% efficiency gain. This value is used as base value for the following variations.

2.3 Variation of compression ratio

It was expected that compression ratio be the main parameter responsible for engine efficiency.

A variation of CR calculated from 16 to 35 showed, that the maximum cylinder pressure rises up to 284 bar at the highest compression ratio. The indicated mean effective pressure (imep) is nearly constant.

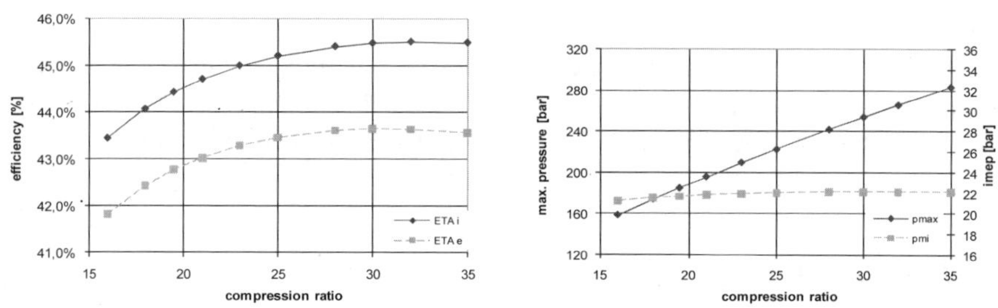

Fig. 4 Effect of compression ratio.

The efficiency gain is great in the CR region between 16 and 25. Higher compression values seem to have no additional benefit. Just in Figure 2 can be seen, that the compression ratio gain gets more flat for higher values. Due to increasing wall heat transfer losses, there is a maximum somewhere between a CR of 28 and 32.

Due to increasing peak pressure, the friction mean effective pressure also increases with higher compression ratios, so that the effective efficiency maximum is reached earlier then the indicated efficiency maximum.

A reasonable maximum value for the compression ratio has been set to ~26.

2.4 A/F ratio variation

The next variation with high potential, read at Figure 2, is the A/F ratio. A more lean mixture has more N_2 and O_2 components and less CO_2 and H_2O components in its combustion gas, which leads to a higher isentropic coefficient. An additional beneficial effect is the lower peak temperature, which may lead to lower heat transfer losses.

Of course, with a more lean mixture, the indicated mean effective pressure drops. This effect would smooth the simulation results and make them not comparable. To avoid this issue, the boost pressure was modified, so that the imep was constant at 21.7 bar.

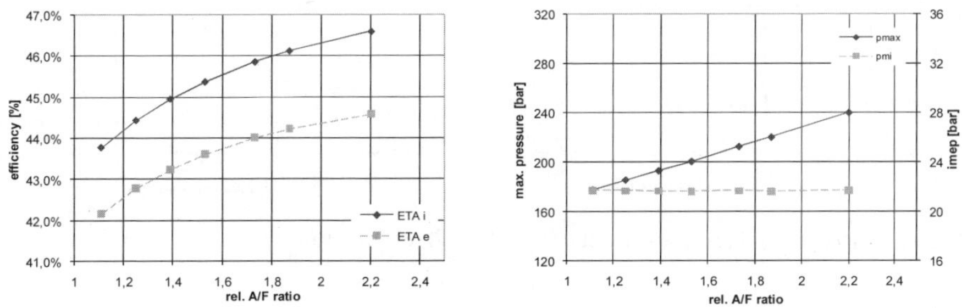

Fig. 5 Variation of rel. A/F ratio with constant imep by raising charge air pressure.

Similar to all other variations, the maximum cylinder pressure also increases with more lean mixture. In this case, it is caused by the higher absolute gas mass due to the increasing boost pressure.

The efficiency gain by lean mixture is enormous. There is a span width of 3%-points in total, while varying the relative A/F ratio from 1.11 to 2.2.

Unfortunately, the extreme lean simulation points are not realistic. The exhaust gas temperature and enthalpy drops, so that there is not enough energy to drive the turbocharger, which has to boost an even higher pressure. The leanest possible relative A/F ratio is in the range of ~1.6.

2.5 Charge air pressure

The last potential can be seen in raising the maximum charge air pressure. There are three motivation factors for raising the charge air pressure.

1. While the indicated mean effective pressure gets higher, the friction mean effective pressure of the engine stays nearly constant or increases slowly. As result, the mechanical efficiency gets better.
2. The energy density increases faster than the losses by wall heat transfer and pumping losses.
3. Considering also part load operation, a part load operating point can be still in the charged area. The efficiency characteristic becomes more flat.

Due to possible cross-effects with A/F ration and compression ratio, we calculated several variations of the charge air pressure, based on compression ratio 19.5 and 26, and relative A/F ratio 1.25 and 1.6.

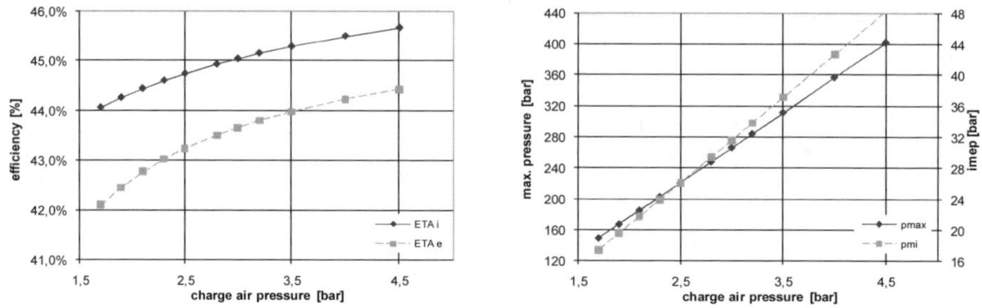

Fig. 6 Variation of boost pressure at CR = 19.5 and rel. A/F ratio = 1.25.

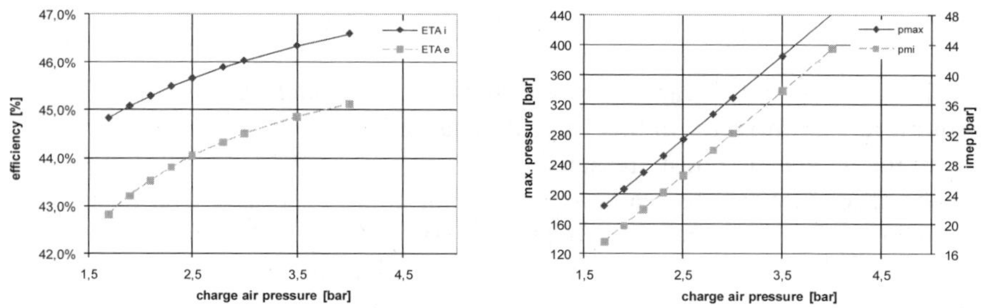

Fig. 7 Variation of boost pressure at CR = 26 and rel. A/F ratio = 1.25.

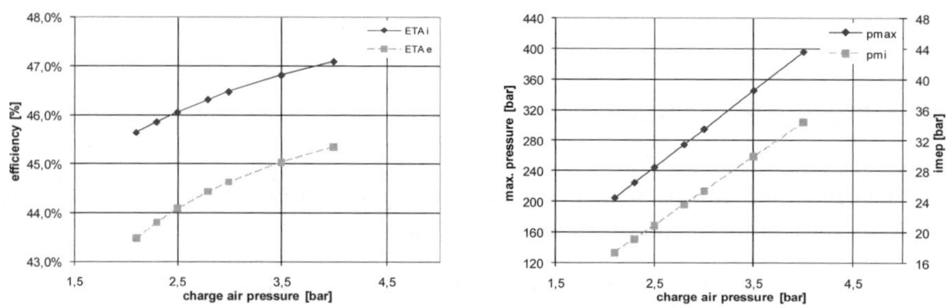

Fig. 8 Variation of boost pressure at CR = 26 and rel. A/F ratio = 1.6.

If we limit the maximum cylinder pressure to 400 bar, the highest potential in imep is given by the base case with CR=19.5 and relative A/F ratio of 1.25. Here the imep value reaches over 48.2 bar, while indicated and effective efficiency raise to 45.6% and 44.4%.

The highest efficiency can be reached with higher CR and more lean mixture. In that extreme scenario with CR = 26 and relative A/F ratio = 1.6, the efficiency raises up to 47.1% and 45.3%. However the power density does not reach such high values, the imep is limited to 34.4 bar here.

2.6 Friction

Until today, it is not possible to give an exact estimation of the friction mean effective pressure of alternative crank drives. However, the potential can be estimated, if the indicated efficiency is used as upper border line.

3 Conclusion

As one can see, there is still some thermodynamic potential left in internal combustion engines, which can not be used formerly due to limitations in maximum cylinder pressure.

The simulation results showed potential of up to 47.1% in indicated efficiency and 45.3% in effective efficiency. This is a further potential of 3% absolute or 7% relative. Additionally, there is some potential with friction-reduced crank drives.

Abbreviations

Effective efficiency	η_e, ETA e
Indicated efficiency	η_i, ETA i
indicated mean effective pressur	imep
break mean effective pressure	bmep
friction mean effective pressure	fmep
Crank angle	CA
Compression ratio	CR

References

Schapiro B. (2008), *The RPM Rotary Piston Machines*. In: Plath, P. J. und Hass, E.-Ch. (Edrs), Vernetzte Wissenschaften, Logos Verlag Berlin 2008.

Schapiro B. and Terlitsky L. (2008), *RKMs: New Class of Machines*. In: Subic A., Leary M. and Wellnitz J. (Edrs), Meeting the Challenges to Sustainable Mobility, Proceedings ICSAT 2008, International Conference of Sustainable Technologies, November 2008, Melbourne, Australia.

Gotter A. (2009), *Softwareprodukte SimEngine und SimThermo*. Available: http://www.gofficient.de. (Accessed: September 28, 2009).
RKM GbR Website: www.rkm-schapiro.org.

FORMULA H – DESIGN AND DEVELOPMENT OF A STUDENT-BUILT HYDROGEN FUELLED RACING CAR

G. R. Pearson[1], T. Wegele[2], A. Subic[1], J. Wellnitz[2]

[1]School of Aerospace Mechanical and Manufacturing Engineering,
RMIT University, Bundoora Victoria 3083 Australia
Email: geoff.pearson@rmit.edu.au

[2]University of Applied Sciences Ingolstadt,
Esplanade 10, 85049 Ingolstadt, Germany,
E-mail: joerg.wellnitz@fh-ingolstadt.de

Abstract Students from RMIT University in Australia and Fachhochschule Ingolstadt in Germany have collaborated to design and build a hydrogen powered racing car. Converting an internal combustion engine to hydrogen fuelling provides a number of challenges, due to the fuel's unique combustion properties. This paper discusses design and implementation hurdles faced by the team, learning outcomes of the project, and demonstrates that environmentally sustainable motoring is achievable with today's technologies.

1. Introduction

Hydrogen has been proposed as a viable energy source for "zero carbon emission" transport. Through reaction with oxygen, hydrogen can be used to produce energy with the only product of reaction being water. Advocates of the "Hydrogen Economy" propose that renewable energy sources such as wind and solar power can be used to produce hydrogen through the electrolytic "cracking" of water (Hoffmann, 2001). Hydrogen can thus be used to store intermittent renewable energy sources, and provide an alternative to fossil fuels that is truly clean and renewable.

A number of automotive manufacturers have been investigating hydrogen as long-term solution to sustainable mobility. There is some split in research preferences, with manufacturers such as Honda and Daimler AG focusing on fuel cell technologies (Hoffmann 2001), whilst BMW and Ford have been investigating hydrogen internal combustion engines (Kiesgen *et al*. 2008, Natkin *et al*. 2007). However there is still some cynicism as to the feasibility of bringing hydrogen powered vehicles to market, leading one commentator to quip "Hydrogen has always been the fuel of the future, and it looks like it always will be" (Strahan, 2008).

J. Wellnitz et al. (eds.), *Sustainable Automotive Technologies 2010*,
DOI 10.1007/978-3-642-10798-6_29,

2. The Formula H Project

2.1 Project objectives

To prove that hydrogen is a feasible option for automotive power, students at RMIT University (Australia) and Fachhochschule Ingolstadt (Germany) have collaborated to design and build a hydrogen powered racing car. Known as "Formula H", the project has engaged over 100 students across the two universities, and has delivered a fully functioning Le Mans prototype style vehicle.

The project's primary objectives have been as follows:

- Prove that hydrogen vehicles are feasible now with existing technologies.
- Promote sustainable automotive technologies to the wider community.
- Give students invaluable exposure to current research and practices in sustainable automotive technologies.

In keeping with the above objectives, the team established the following guiding principles for its operations:

- All design and manufacturing work must be completed by students, to prove that hydrogen is not only the domain of highly trained specialists.
- Design choices must be simple, commercially feasible and based on existing automotive technologies.
- The final product must be visually exciting and build enthusiasm for green motoring.

2.2 Project structure

The project team had a timeframe of approximately 18 months from inception to the vehicle's launch in Melbourne at ICSAT 2008. The project was broken down as follows:

- The RMIT University team would design and manufacture the rolling chassis and bodywork, capitalizing on the in-house vehicle design knowledge gained through the university's successful Formula SAE program.
- The Fachhochschule Ingolstadt team would develop the powertrain and fuelling system on their own test stand, with guidance from local experts in hydrogen automotive systems such as TÜUV SÜD.
- After approximately 12 months, the powertrain system would be shipped to RMIT, where students from both teams would work together to complete the vehicle.

3 Vehicle Conceptual Design

With regard to the project objectives outlined above, the following vehicle-level design decisions were made at the commencement of the project:

- **Hydrogen Internal Combustion Engine (H_2ICE)**: The H_2ICE option was chosen over a hydrogen fuel cell due to advantages in cost and power density. It also represents an option that can be implemented in the near future with existing supply chain infrastructure, which agrees with the overall objectives of the project.
- **Steel space-frame chassis construction**: This was considered to be the most practical means to build a cheap, light and stiff chassis structure. Compared to more advanced methods such as carbon monocoque construction, a simple steel structure can be easily modified which was considered critical for a first attempt at a hydrogen vehicle.
- **Commercially available unsprung components**: Use of Formula Ford wheels, hubs, brakes and uprights limited time spent on non-productive design tasks.
- **LeMans prototype styling**: This would create a dramatic looking vehicle without drawing comparison to existing road vehicles or other university projects.

4 Vehicle Detail Design

4.1 Vehicle architecture

The vehicle is rear engined, with the driver sitting along the centerline of the vehicle. A core design goal was to isolate the hydrogen system from potential ignition sources. As such, the entire hydrogen storage and supply system is contained in the left hand sidepod of the vehicle, whilst the right hand sidepod contains the majority of the electrical system, including battery, engine management system and wiring loom. The tank compartment is ventilated to atmosphere through an air intake at the front and a vent at the rear, to eliminate build up of trapped gases, and each compartment of the vehicle is isolated with aluminium panels fitted snugly to the spaceframe.

The offset location of the heavy hydrogen tank in the left hand sidepod creates a slight mass imbalance across the vehicle. To counter this, batteries and radiator are positioned in the right hand sidepod, but calculations indicate the centre of gravity of the vehicle to be approximately 50mm to the left of the longitudinal centerline. This was considered an acceptable trade-off given the abovementioned safety considerations.

Fig.1 Formula H vehicle. Note hydrogen tank visible in left hand sidepod.

4.2 Engine

The engine chosen was a Rotax 800cc parallel twin cylinder motorcycle engine, featuring balancing conrod, 4 valves per head and double overhead camshafts. On petrol fuel, the maximum engine output is 62.5kW (8000rpm), and maximum torque is 86Nm (5800rpm). The twin cylinder engine is a simple configuration that provides reasonable performance while limiting the number of components required for hydrogen conversion.

Hydrogen combusts with much higher flame speeds and a smaller quenching zone than petrol, and this can lead to much higher thermal loads on combustion chamber components (Kiesgen *et al.* 2006). Presently engine internals have been left standard, but this may be reviewed if thermal failures become apparent.

Hydrogen also has a wide range of flammability, and can combust in a mixture range of 4-75% in air. Maximum power output would be achieved at an air-fuel ratio of around lambda = 1, but at this ratio there can be significant abnormal combustion. To minimize potential engine damage, the team is running the engine lean, with a minimum lambda value set at 1.3. This lean combustion, when combined with the oxygen displacement effect of port fuel injection, results in expected maximum torque and power outputs approximately 60% of the base petrol fuelled engine.

4.3 Hydrogen supply system

Hydrogen fuel is stored onboard in a class three carbon fibre filament wound tank with a seamless aluminium liner. The tank has a capacity of 72 litres, has a weight of 32kg, is rated to 20Mpa, and when fully charged holds 1kg of hydrogen.

A schematic diagram of the hydrogen supply system is provided in Figure 2. Features include manual and automatic high-pressure isolation valves adjacent to the tank, pressure reduction down to 500kPa bar via a single mechanical regulator, and delivery to the port-mounted Hoerbiger injectors via flexible line. Hydrogen compatible fittings are used throughout, and team members have received training and certification to assemble all fittings.

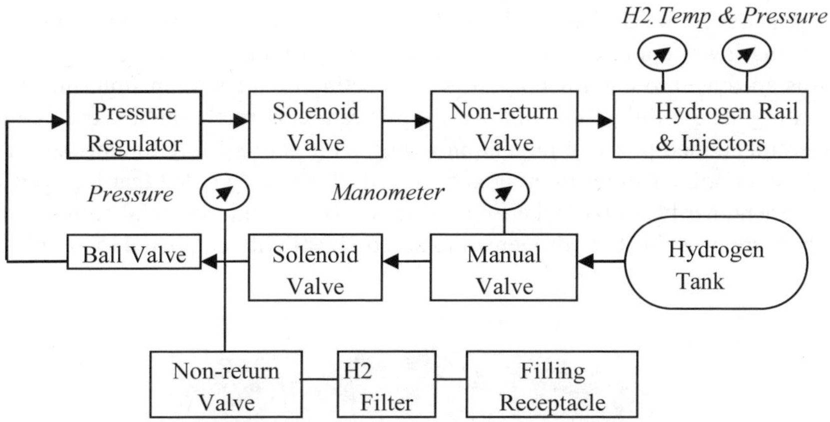

Fig. 2 Hydrogen storage and supply system schematic diagram.

5 Testing and Development

5.1 Pressure regulation

Early tests revealed very poor running and significant amounts of abnormal combustion. This was traced to irregularities in the fuel rail pressure, which was varying erratically between 100kPa and 500kPa. A mechanical regulator reacts to the pressure difference across the regulator, and when fuel demand varies with changing engine load the regulator adjusts accordingly. However the adjustment occurs with some time delay, and this delay was causing the pressure variation. With inconsistent pressure it is impossible to finely tune injection pulse-widths, and the variation in air-fuel ratio was causing the irregular combustion, which exacerbated the problem. After experimenting with in-line buffer reservoirs, a satisfactory short-term solution was found by fitting a larger regulator.

The abovementioned dynamic effect is typical of mechanical regulators, and can be accounted for with a fuel rail pressure compensation tuned into the engine management system. However for better performance the team is investigating more satisfactory solutions such as multi-stage or electronic regulation.

5.2 Piston Failure

The engine experienced piston failure during early testing, whereby the ring land between the lower compression and oil control rings broke away, (see Figure 3). This was attributed to unburned hydrogen collecting between the piston rings and self igniting. The failure coincided with significant abnormal combustion, due to the abovementioned pressure regulation issues, so it is premature to consider piston redesign until smoother running is achieved. It was duly noted that Kiesgen *et al* recommend hard anodizing the ring grooves, to cope with potential combustion in the ring crevices due to hydrogen's higher combustibility compared to petrol.

Fig. 3 Failed piston. Note broken ring land segment between compression and oil ring.

5.3 Dynamometer tuning

Initial chassis dynamometer tuning at RMIT University was unrepresentative due to the abovementioned pressure regulation and abnormal combustion issues. Maximum power achieved during RMIT testing was 26kW, although the team is confident it has now surpassed this figure with the resolution of the fuel pressure issues. The vehicle has since been transported to Germany where it is presently undergoing drive testing and promotional activities.

The vehicle will be returned to RMIT in early 2010, where dynamometer tuning will resume and calibration will continue to be fine-tuned.

6 Future Work and Conclusions

The base vehicle manufacture is now complete, and the team is moving into the developmental stages of the project. Progress has been slowed by difficulty in sourcing appropriate testing venues, but the team is working to the following development plan:

1. **Calibration**: The team will continue to refine injection and ignition maps for the existing engine, in order to establish the engine's performance "baseline".
2. **Modification**: Pending results of the calibration testing, the team will assess internal modifications required to eliminate any hydrogen combustion irregularities. This will include review of compression ratio and piston design as outlined previously
3. **Performance enhancement**: Once the engine is running smoothly, a series of performance enhancements are planned to increase power density. This will include investigations into electric supercharging and cryogenic cooling of the intake charge.
4. **Hydrogen production**: RMIT University has significant research experience in production of hydrogen using solar and wind energy. A long term plan is to demonstrate fully sustainable motoring, whereby hydrogen produced onsite using renewable means is used to power the Formula H vehicle.

Despite early setbacks the team has managed to deliver a fully functioning hydrogen powered racing car, in accordance with the stated project objectives. As a student learning exercise it has exposed the student team members to the full spectrum of the sustainability issues facing the automotive industry, from focussed engineering problem-solving through to wider issues such as fuel supply and infrastructure, risk management, global co-operation and addressing varying public sentiment. The team hopes that its efforts are proving that environmentally sustainable motoring is an achievable goal.

References

Kiesgen G (2006) The New 12-Cylinder Engine in the 7 Series: The H2 ICE Age has begun, *2006 SAE World Congress*, Detroit Michigan, April 3-6, 2006. SAE International, Warrandale, PA.

Hoffmann P (2001*) Tomorrow's Energy: Hydrogen, Fuel Cells and the Prospects For a Cleaner Planet*, MIT Press, Cambridge, Massachusetts.

Natkin R J , Denlinger A , Younkins M, Weimer A.L.Z., Hashemi S., and Vaught A.T. (2007) Ford 6.8L hydrogen IC engine for the E-450 shuttle van, *Powertrain and Fluid Systems Conference and Exhibition*, Chicago Illinois, October 2007, SAE International, Warrandale, PA.

Strahan D (2008) Hydrogen's long road to nowhere, *New Scientist*, November 29 2008, 40-43.

FORMULA H – RISK ASSESSMENT AND SAFETY STRATEGIES IN A STUDENT HYDROGEN VEHICLE PROJECT

G. R. Pearson[1], S. J. Simmons[1], P. Mehringer[2], A. Subic[1], J. Wellnitz[2]

[1]School of Aerospace Mechanical and Manufacturing Engineering,
RMIT University, Bundoora Victoria 3083 Australia
Email: geoff.pearson@rmit.edu.au

[2]University of Applied Sciences Ingolstadt, Esplanade 10, 85049 Ingolstadt, Germany
E-mail: joerg.wellnitz@fh-ingolstadt.de

Abstract Students from RMIT University and Fachhochschule Ingolstadt have manufactured a hydrogen powered racing car. Due to a lack of globally recognized design standards, the team has adapted existing gas vehicle standards and proposed hydrogen draft regulations, and performed a comprehensive Failure Mode Effect Analysis to establish project safety protocols. This paper reports on the team's experiences and findings.

1. Introduction

A transportation system based on hydrogen fuel offers the potential of sustainable and carbon-free transport, an urgent imperative given the consequences of anthropogenic carbon dioxide emissions and diminishing fossil fuel reserves. The automotive industry is facing a considerable shift in consumer sentiment towards sustainable automotive technologies. Hydrogen power is one such technology, and a number of automotive manufacturers are looking to hydrogen as a long-term energy supply solution.

RMIT University and Fachhochschule Ingolstadt have collaborated to design and build a hydrogen-fuelled racing vehicle. Known as *Formula H,* the project has delivered a fully functioning hydrogen powered Le Mans prototype style vehicle. The RMIT team has designed and manufactured the rolling chassis, whilst the Fachhochschule Ingolstadt team has designed and developed the powertrain package. Integration of the two systems has been a combined responsibility between the two teams.

J. Wellnitz et al. (eds.), *Sustainable Automotive Technologies 2010*,
DOI 10.1007/978-3-642-10798-6_30,

Fig. 1 The RMIT University / Fachhochschule Ingolstadt Formula H vehicle.

The primary purpose of the project has been to demonstrate that hydrogen is a viable and safe option as a sustainable automotive fuel, and can be implemented using existing technologies. Thus, a key feature of this project has been the use of a solely student workforce to design, manufacture and develop the vehicle. The team hopes to prove that if a relatively inexperienced team with limited resources can successfully build a hydrogen vehicle, then it should be possible for major manufacturers to do the same. It is also hoped that successful demonstration of the final vehicle will help allay some public perception that hydrogen is a dangerous fuel. Anecdotal evidence gathered by the project team at promotional displays of the vehicle has shown that a significant proportion of the populace still link hydrogen fuel to catastrophe, through exposure to incidents such as the Hindenburg disaster or the use of the hydrogen bomb. Such a perception can only be overcome through repeated, demonstrated safe usage of the fuel.

2. Hydrogen Vehicle Safety Strategies

2.1 Existing regulations and safety advice

A significant issue faced by the project team has been the sourcing of an appropriate set of design standards specifically for the manufacture of hydrogen vehicles. Whilst an abundance of general design advice exists, there is little to be found in the way of unified regulatory standards on which to base a whole vehicle approval. This has created problems for the project team, as organizing testing venues and public demonstrations of the vehicle has been very difficult whilst the vehicle lacks some form of signed regulatory approval. This then becomes a barrier to the team's stated goal to publicly demonstrate safe usage of hydrogen fuel.

The Formula H project team has taken leadership by using available information on hydrogen and other gaseous fuels to develop their own set of internal protocols to manage the associated risks.

Given the team would be using commercially available and approved components for the entire hydrogen storage and fuel delivery system, it was not deemed necessary to perform detailed risk analysis at an individual component level. The issue facing the team was assessing the safety of the vehicle at a systems level, covering interdependencies between components, integration of the hydrogen system into the complete vehicle, and the development of vehicle operating and testing procedures. Initial investigation of hydrogen risk management and safety advice at a vehicle system level revealed a range of reference material, including:

- UNECE draft regulations for hydrogen vehicles (UNECE, 2004).
- SAE Technical Information Reports (SAE, 2008).
- Material Safety Data Sheets (MSDS) from hydrogen suppliers.
- General hydrogen safety advice from bodies such as the National Hydrogen Association.

It was noted that none of the above were legally binding documents. The UNECE document was a proposed draft guideline that was yet to be approved, and whilst the SAE standards provided useful guidance they stated that vehicle approvals need to be addressed within each country. MSDS documents provided useful info on hydrogen handling that could be used to develop operating procedures.

2.2 Formula H project safety strategy

Given the lack of binding regulations, the team undertook a safety strategy that integrated advice from a number of sources. The Fachhochschule Ingolstadt team was tasked with powertrain design and development, and implemented the following strategy:

- Use UNECE draft regulations as broad guidelines for overall system design.
- Consult with TÜV SÜD for advice on selection of individual components, system design and operational procedures.

The RMIT team was tasked with rolling chassis design and development, and implemented the following strategy:

- Use existing Australian Standards for Compressed Natural Gas fuelled vehicles, (AS/NZS 2739) as the basis for chassis design decisions concerning the gaseous fuel system.

- Design the vehicle roll hoop and driver protection structure in accordance with the regulations of the Confederation of Australian Motor Sport, Australia's governing body for motorsport (CAMS, 2008).

The project team believed the above strategy to be the most practical means by which to progress, given lack of unified global guidelines and the difficulties of designing a vehicle across two continents. The CNG vehicle standards were considered appropriate for a hydrogen vehicle given CNG's similar installed system pressures and given that it shares hydrogen's tendency to rise in air.

2.3 Comparison of hydrogen and CNG safety guidelines

It is beyond the scope of this report to provide detail of individual clauses, but comparison of the hydrogen vehicle draft regulations (UNECE, 2004) and CNG vehicle standards (Standards Australia, 2003) revealed a number of common and significant threads, including:

- Strong emphasis on isolating potential gas leaks from possible ignition sources, such as electrical items or hot exhaust manifolds.
- Ventilation requirements to prevent escaped gas from collecting in pockets around the car.
- Requirements for protection of the complete gas supply system including all lines and fittings.
- Standards for refueling fittings and design integrity of the surrounding bodywork
- Requirements for various manual isolation and automatic fuel shutoff valves adjacent to the gas cylinder.

It was noted that the Australian CNG standards only allowed a maximum system pressure of 20MPa, whereas European hydrogen standards allow system pressures up to 70MPa. However given the vehicle was designed to use a 20MPa tank this was not considered to be an issue.

3 Formula H Vehicle Design

With regard to the abovementioned recommendations, the Formula H vehicle has been designed with a "cold side" and a "hot side" along either side of the vehicle, separated by the central driver compartment and engine bay (Figure 2). The "cold side" runs along the left hand side of the vehicle, and contains the complete hydrogen system including tank, refuelling dock, and the various lines and fittings that comprise the hydrogen fuelling system. This feeds hydrogen to the engine through a single low pressure line entering the engine bay from the upper left hand side. The "hot side" on the right hand side of the vehicle contains the majority of heat and potential ignition sources such as the cooling system and fans, batteries, wiring loom, relays and engine management system. The various compartments are isolated from each other through aluminium panels fitted snugly to the space-frame, and both left and right hand sides are ducted to provide free air flow through the compartments for ventilation to atmosphere. Any leaked hydrogen gas is detected through a hydrogen sensor placed in the tank compartment, which when activated instantaneously cuts off ignition via the engine management system.

The electronic components nearest to the hydrogen compartment are the dashboard switches and loom in the central driver compartment. Special attention was paid to sourcing spark-free switches, and sealing all wiring inside a gas-tight dash panel.

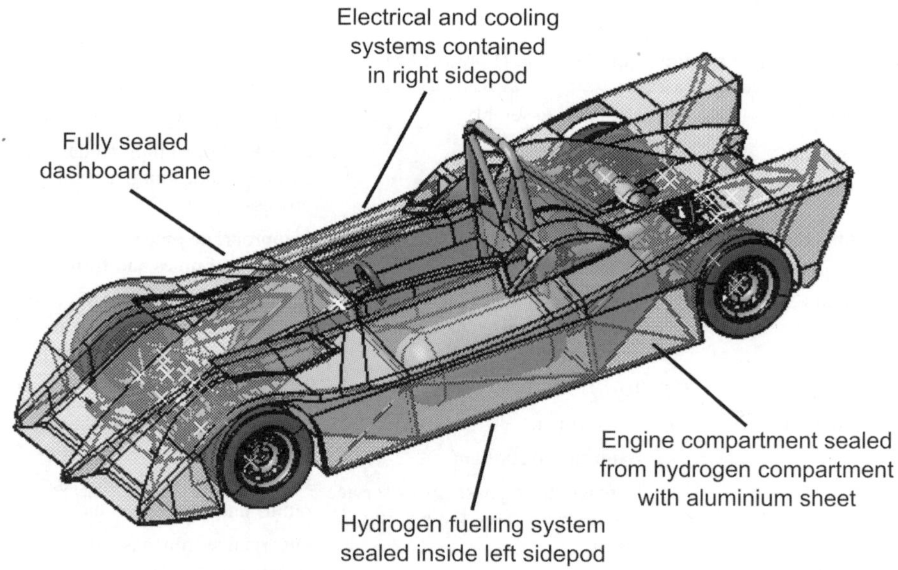

Fig. 2 Formula H vehicle architecture.

The rear engine bay necessarily requires co-location of hydrogen fuel lines with potential heat and ignition sources, so steps were taken to separate these as far as practical. The exhaust pipe is routed low and to the right hand side of the engine, whilst the hydrogen delivery lines reside in the upper left hand side of the engine bay, mostly above the engine. Any leaked gasses should rise to the top of the engine cover where they are safely vented to atmosphere.

4 Failure Modes Effect Analysis

A comprehensive Failure Modes Effect Analysis (FMEA) study was undertaken by the team to identify major risk factors in the vehicle design and the team's operational procedures. The team has systematically worked its way through possible failure points and rated them according to the severity of a failure, the likelihood of the failure occurrence, and the likelihood of detection. The product of these three scores gives a Risk Priority Number that ranks the overall risk, with high risk scores requiring additional actions or particular attention.

Samples of high and low scoring risk factors are given in the FMEA excerpt shown in Table 1 below. Note that with the Tank Assembly, additional actions are advised in the regular inspections and leak-testing of the tank after service. This advice was subsequently incorporated into the team's operating procedures. In the case of the hydrogen pipes and fittings, identifying the potential safety hazard led to accredited staff training in pipe and fitting preparation, which therefore minimized the overall risk factor.

Table 1 Sample excerpts from Formula H project FMEA.

System	1.1 Tank Assembly	2.1 Hydrogen pipes and fittings
Sub-System	1.1.1 Tank Surface	2.1.1 Hydrogen pipe preparation: cutting and bending, assembly process
Potential Failure Mode	Damage of Tank Surface	Inappropriate pipe preparation / assembly if pipes and fittings
Potential Failure Effects	Cracks in tank surface affecting stiffness of tank; tank burst in worst case	Re-assembly of pipes not possible; installation safety not assured; fatigue, leakage, pipe burst
SEVERITY	10	2
Potential Causes	Stone Impact Exposure to sunlight Damage during assembly / service, impact from parts, tools during testing	Pipe cutting: cutting force too high Pipe bending: minimum bending radius under-run Divergence from assembly instructions
OCCURENCE	3	5
Current Process Controls	Tank compartment construction ensuring protection against external influences, including underbody construction (Al-honeycomb)	Course of instruction with components supplier (HPS) conducted in Feb 2008 in Ingolstadt with certification of participants Assembly work on fittings restricted to certified team members Perform tightness testing according to service instruction
DETECTION	4	2

RISK PRIORITY NUMBER	120	20
Actions recommended	Regular leakage tests Regular visual inspections	

5 Further Work

The vehicle is still in its early stages of development, and additional work is planned to bring it to full maturity. The team has proven the integrity of the hydrogen system itself, and attention will now be focussed on refinement of the tank compartment including:

- Refinement and gas flow testing of the tank and adjacent compartments to ensure gas-tightness under potential leak conditions.
- Physical testing of the chassis tank protection structure.

Until such tests are completed, drive testing will only take place under strictly limited conditions, specifically with an open tank compartment to eliminate any gas entrapment.

Further to the above testing, the team is continuing to refine the project FMEA and implementing resultant operating procedures.

6 Findings and Recommendations

Despite the team's best efforts to comprehensively research, implement and document thorough safety procedures, the lack of unified standards and approval structures have caused significant problems. Failure to obtain some form of regulatory approval for the vehicle has resulted in difficulties in accessing testing venues, and has resulted in at least one major promotional event for the team being cancelled. Research into hydrogen as an automotive fuel must be encouraged at all levels if it is to be implemented, but the present situation provides hurdles that can be difficult for a small project team to overcome. The Formula H team strongly urges the automotive hydrogen community to continue pushing forward to ratify a unified set of design standards for hydrogen vehicles.

More positively, engaging team members in the research and development of their own safety procedures has strengthened their awareness of the risks involved, and given them a distinct sense of ownership of the final processes. For what is essentially a student project, this is a great learning outcome.

References

CAMS (2008), General Requirements for Cars and Drivers, *2008 CAMS Manual of Motorsport,* Confederation of Australian Motorsport, Malvern East, Australia.

SAE (2008), Technical Information Report for Fuel Systems in Fuel Cell and Other Hydrogen Vehicles, SAE Standard J2579, Society of Automotive Engineers International, Warrandale PA.

Standards Australia (2003), Natural gas (CNG) fuel systems for vehicle engines, Australia New Zealand Standard AS/NZS 2739:2003, published 2003.

UNECE (2004), *Proposal for a new draft regulation: Uniform provisions concerning the approval of 1. Specific Components Of Motor Vehicles Using Compressed Gaseous Hydrogen; 2. Vehicles With Regard to the Installation of Specific Components for The Use Of Compressed Gaseous Hydrogen,* Geneva (online),
Available: http://www.unece.org/trans/main/ wp29/wp29wgs/wp29grpe/grpedoc_2004.html. (Accessed 25 September 2009)

Part 7
End-of-life vehicle management

INCREASE OF RECUPERATION IN VEHICLES WITH CONVENTIONAL POWERTRAIN

L. Brabetz [1], M. Ayeb [1], D. Tellmann [1]

[1]Department of Electrical Engineering and Computer Science, University of Kassel, Germany, Wilhelmshoeher Allee 73, 34121 Kassel, Germany,
E-mail: brabetz@uni-kassel.de, ayeb@uni-kassel.de, tellmann@uni-kassel.de

Abstract A well known approach to enhance the energy efficiency of a vehicle is the recuperation of its kinetic or potential energy during negative acceleration or downhill driving. Electrical or hybrid powertrains are well suited for this concept, but current vehicles with IC engines can use it, if at all, only on a very limited level. On the other hand, even aggressive prognoses expect the majority of future vehicles to be based on IC engines. The presented concept uses recuperation to supply the energy for the electrical distribution system. This approach allows a significant efficiency improvement as the electrical power consumption in future vehicles is expected to increase considerably. It is based on a modified regular alternator in combination with a DC/DC converter, ultra-capacitors and a specific energy management system. A hardware in the loop based test set up has been built in order to validate the described approach. Besides the theoretical description and modelling, the paper presents first experimental data and results followed by a discussion of future potentials.

1. Introduction

In order to implement efficient recuperation in conventional vehicles, three conditions are to be fulfilled: (1) the power of the energy converter, e.g. the alternator, must be high enough, (2) the electrical storage device must be properly dimensioned and (3) the electrical power consumption of the vehicle should allow the use of the regained energy. Unfortunately, the first two conditions have negative design impacts on packaging, weight and costs, whereas the third one will already be valid in the near future.

Initial vehicles already in production are using standard or improved alternators and batteries in combination with an energy management which takes into account the current driving condition. The downside of this concept is that the alternator operates on battery voltage level and that the frequency and depth of charge cycles is increased. As a consequence, the maximum power of the alternator is limited and the lifetime of the battery may be shortened.

Other approaches address these disadvantages by adding ultra capacitors to handle the critical charge cycles and by operating the alternator at higher voltages.

J. Wellnitz et al. (eds.), *Sustainable Automotive Technologies 2010*,
DOI 10.1007/978-3-642-10798-6_31,

These concepts require DC/DC converters and optional switches and cause negative design impacts on packaging, weight and costs. On the other hand, they provide a higher recuperation rate, an improved stabilization of the supply voltage and downsizing of the battery.

Different architectures were realized or have been proposed. The one described in this paper is designed to minimize the power requirements to be met by alternator, electrical storage devices and converter, thus reducing the cost impact under the boundary condition of improved energy efficiency. It is based on a voltage controlled alternator and an appropriate charging strategy for a combination of super capacitors and battery.

2. Architecture

The architecture depends on the functional requirements such as

- sufficient and robust energy supply of all electrical loads
- stabilization of the supply voltage
- additional supply voltage and its stabilization for specific power loads
- improved power characteristics to handle short time high power requirements
- enhanced start capabilities even at low SoC
- and energy efficiency and recuperation capability.

Furthermore, the architecture is determined by the technical and economical targets and boundary conditions, which determine the parameters of the applied components:

- type of starter (conventional/start/stop), alternator, battery and voltage control
- use and performance of an additional storage device and of DC/DC converters
- power loads and their voltage range
- and energy management and switching devices.

As a consequence, there are a high number of possible architectures and corresponding topologies of which some are already in production or are under investigation, for example [1-6].

The first approach is to obtain maximum use and functionality by employing regular or improved standard batteries and by applying minimum design changes to a current architecture as shown in Figure 1a. The solution as in Figure 1b allows the realization of a low cost start-stop function via a DC/DC converter and the architecture presented in Figure 1c utilizes the voltage control of the alternator G_2 to render possible fast charging of the battery during vehicle braking and thus to allow a moderate recuperation. The optional DC/DC converter stabilizes the voltage supply of sensitive electrical components.

The limitations of current batteries can be overcome either by applying a more innovative technology such as LiIon or by adding a second energy storage with complementary technical characteristics as provided by ultra capacitors (Figure

2a). A solution as shown in Figure 2b applies a DC/DC converter to the ultra-capacitor and thus makes a better use of its energy content due to $E=CU^2/2$. Alternatively, it can be used to provide a power supply with a higher voltage and a low internal resistance.

Within these architectures, an efficient recuperation strategy is limited by the maximum power and current of the alternator and the battery. An obvious approach is to let the alternator charge the ultra-capacitor by utilizing an adapted voltage control (Figure 3a).

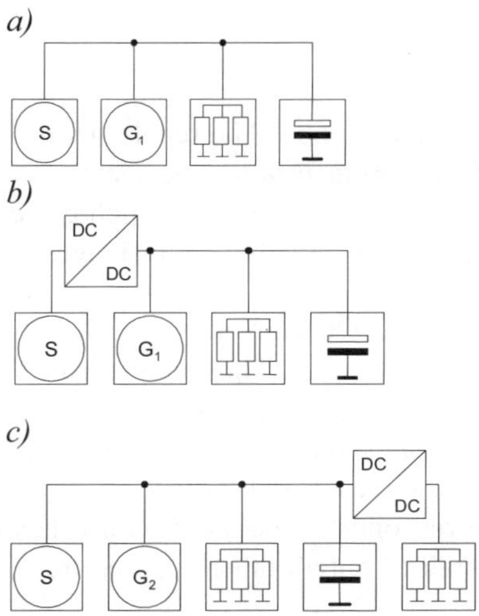

Fig. 1 a, b and c, see text.

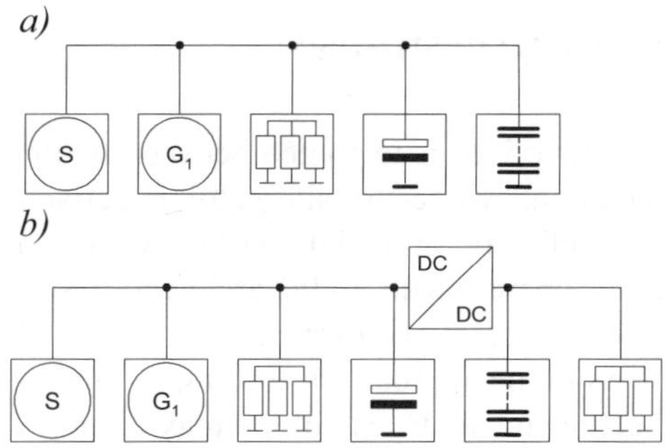

Fig. 2 a and b. see text.

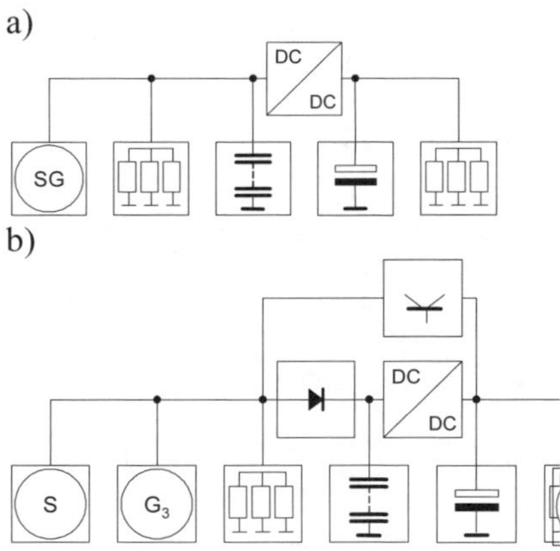

Fig. 3 a and b, see text.

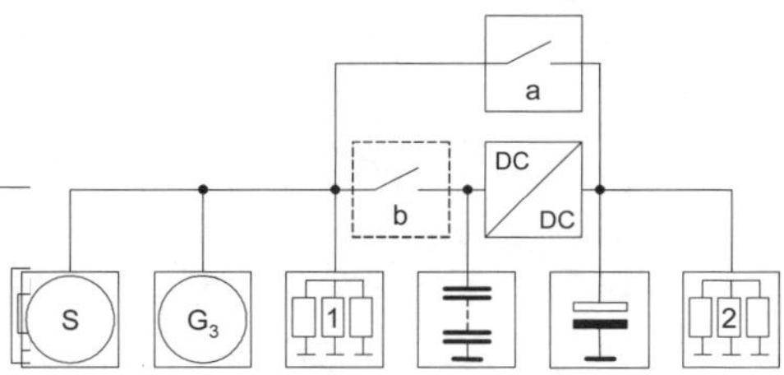

Fig. 4 proposed architecture.

At higher voltages the possible alternator power output increases considerably, its energy efficiency improves slightly and the low internal resistance of the ultra-capacitors allows the required charging currents. Considering all operating states of the vehicle, this converter has to be specified for strong energy flows, which is certainly a drawback. An active cooled linear power switch and a diode as shown Figure 3b could be used to keep the requirements to the DC/DC converter within an acceptable range. The architecture discussed in this paper (Figure 4) is based on the type shown in Figure 3 a, b but uses two relays (or optionally just one, see below) which leads to a very different type of control strategy which will be described in detail in the following section [10].

3. Control strategy

The basic idea is to switch between two modes of operation. The regular mode of operation, where the alternator is active mainly during negative acceleration or downhill driving, is defined by switch *a* open and switch *b* closed, which corresponds operationally to the architecture shown in Figure 3a with all advantages and functions described there.

As in the approach according to Figure 3b, the DC/DC converter is designed to meet the mean instead of the maximum power requirements. Thus on one hand the converter costs are reduced, but on the other hand scenarios are possible where the required power exceeds the maximum rating $p_{DC/DC}$ of the DC/DC converter [10]. In such cases the second operating mode (switch *a* closed and switch *b* open) has to be applied (Figure5). This configuration corresponds to the architecture shown in Figure 2b. In this mode, recuperation is restrained but all other features are still operative.

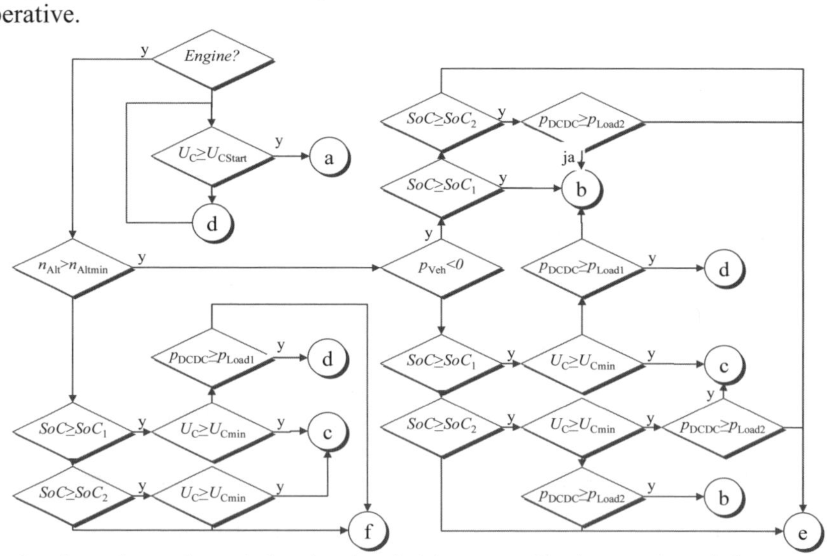

Fig. 5 a-d: regular mode, switch a closed, switch b open., **e-f** back up mode, switch a open, switch b closed, n_{Alt}, n_{Altmin}: rotational alternator speed/minimum, U_C, U_{cmin}, U_{CStart}: capacitor voltages (minimum, minimum.start voltage), SoC, SoC_1, SoC_2: battery state of charge/minimum/minimum1 & 2, p_{DCDC}, p_{Load1}, p_{Load2}, p_{Veh}:max.power of the DC/DC converter, power of loads1 & 2, recuperation power

Switch *b* is optional. If it is not applied, only the second mode is affected and corresponds to the one of the architecture shown in Figure 2a. The only disadvantage would be that the ultra capacitors would be used only partially for energy storage.

A major challenge is a smooth transition from one operating mode to the other without affecting the supply voltage too much and to enable the use of simple switches. This strategy increases the complexity of the voltage control considerably.

4. Measurement set-up

To prove the feasibility of the concept a measurement set-up as shown in Figure 6 was built. It was also used to get the extended technical characteristics of the alternator and the ultra-capacitor string.

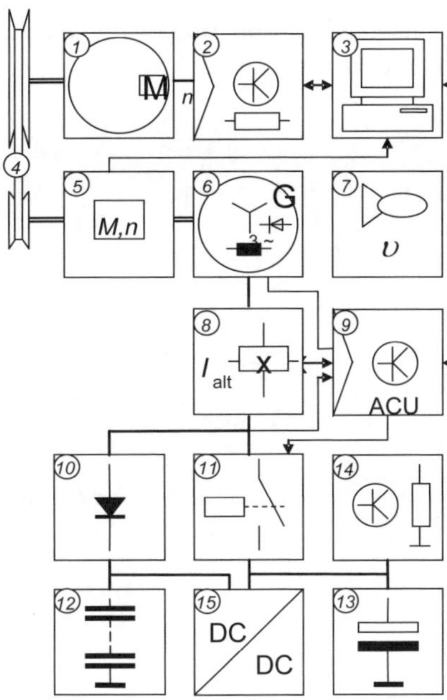

1. 15kW synchronous machine,
2. machine control unit
3. control and data acquisition system
4. regular alternator belt
5. sensor frequency and torque
6. modified 140A-alternator
7. stand-alone thermal camera system
8. sensor alternator voltage and current
9. alternator control unit
10. power Schottky Diode (switch b)
11. relay (switch a)
12. 300F/ 27V ultra-capacitor string
13. battery
14. electronic loads
15. DC/DC converter

Fig. 6 Measurement Set-up.

5. Energy efficiency

Alternator: At a higher operating voltage the alternator is capable of supplying increased power but has to deal with the corresponding augmented absolute heat loss. As to be expected, the efficiency ratio η_{alt} improves slightly with the operating voltage as the output power increases and is shown as a function of rotational speed and alternator current in Figure 7 for different voltages. The measurements of temperature vs. time at the stator coil as the most critical part show a considerable thermal inertia allowing an overload for a limited time as it is the case during recuperation. However, this is still to be verified by appropriate endurance tests.

Ultra-Capacitor: A real ultra capacitor can be described [7] by an equivalent circuit involving additional resistors and capacitances. Charging procedures as shown in Figure 8 (alternator current I_{alt}, voltage U_C and the accumulated charge Q_C versus time t) are used to calculate the efficiency ratio η_C of the charging process. One obtains η_C>95% which exceeds the performance of a battery with only marginal aging effects.
As to be expected, the switching affects a considerable drop of the battery voltage. Hence a support current by the DC/DC converter is mandatory.

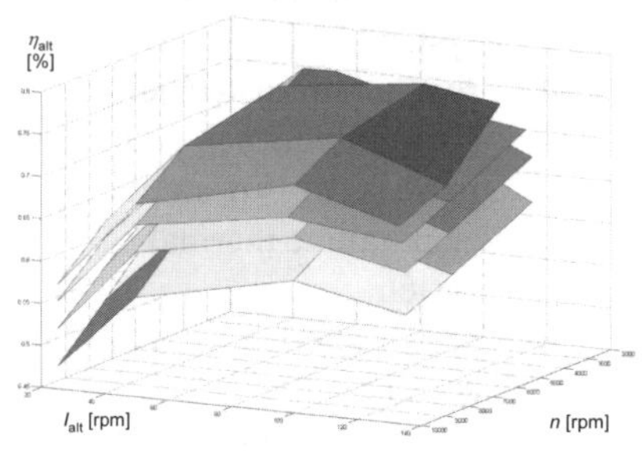

Fig. 7 Efficiency ratio of alternator: U_{alt}=15V,19V, 23V, 26V.

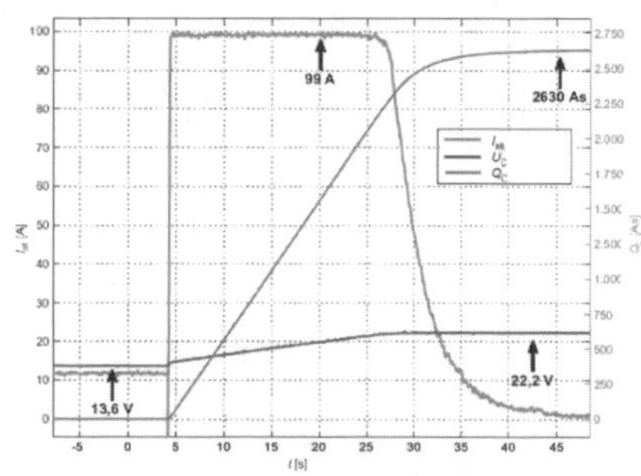

Fig. 8 Charging procedure after switching.

6. System considerations

The capacitance C and its maximum capacitor voltage U_{Cmax} have a considerable impact on both the system cost and efficiency. If C_{cell} and U_{cell} are the capacitance and the voltage of a single capacitor, k the total number of capacitors and n, m the numbers of capacitors in parallel and in series respectively, one gets:

$$k = mn \quad , \quad U_{C\max} = mU_{cell} \quad , \quad C = C_{cell}\frac{n}{m} = C_{cell}k\frac{U_{cell}^2}{U_{C\max}^2}$$

The maximum energy E_{rec} gained during a recuperation cycle of length t_{rec} is given by:

$$E_{rec} = \int_0^{t_{rec}} U_C I_{alt\_max} dt = \begin{cases} I_{alt\_max} U_{C\min} t_{rec} + \dfrac{U_{C\max}^2 I_{alt\_max}^2 t_{rec}^2}{2U_{cell}^2 C_{cell} k}, & t_{rec} < t_{\max} \\ \dfrac{kC_{cell} U_{cell}^2 \left(U_{C\max}^2 - U_{C\min}^2\right)}{2U_{\max\_C}^2}, & t_{rec} \geq t_{\max} \end{cases}$$

$$t_{\max} = \frac{U_{cell}^2 C_{cell} k}{U_{C\max}^2 I_{alt\_max}} \left(U_{C\max} - U_{C\min}\right)$$

where $U_{C\text{min}}$ is the capacitor-voltage at the beginning of the recuperation and I_{alt_max} is the maximum alternator current. Hence $U_{C\text{max}}$ must be as high as possible in order to maximize E_{rec}. Taking into account the maximum voltages $U_{Alt\text{max}}$ and $U_{Sys\text{max}}$ of the alternator and the system, one gets $U_{C\text{max}} = \min(U_{Alt\text{max}}, U_{Sys\text{max}})$.

On one hand $U_{C\text{min}}$ should be as high as possible to minimize the range of the supply voltage of the loads 1 (Figure 4) and to increase the minimum recuperation power $I_{\text{alt}} U_{C\text{min}}$. On the other hand, the maximum recuperation energy decreases with a higher $U_{C\text{min}}$.

Due to the various input parameters and scenarios, the optimum value kC_{cell} is best determined by a simulation approach. Based on the measured characteristics of the components and the modes of operation described in Figure 5, a model [10] has been used to calculate the potential energy saving. It takes into account the influence of the drive cycle [8], simplified vehicle controls and dynamics, fuel consumption relationships, el. power consumption and various system parameters.

7. Measurements and results

A simulation based on a FTP75 drive cycle and a lower middle class vehicle was used to investigate different scenarios of C, $U_{C\text{min}}$ and I_{load} (which was assumed to be constant). As shown in Figure9, there is an optimum $U_{C\text{min}}$ depending on C. Another result is, that there is no significant gain of fuel saving by increasing C above a given value. As discussed, the range $U_{C\text{min}} - U_{C\text{max}}$ increases with smaller values of C (Figure10).

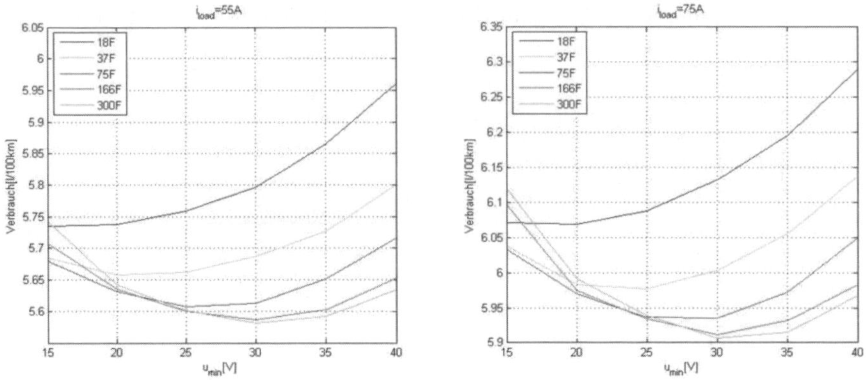

Fig. 9 Average fuel consumption over drive cycle versus U_{Cmin} for various C and $I_{Load.}$

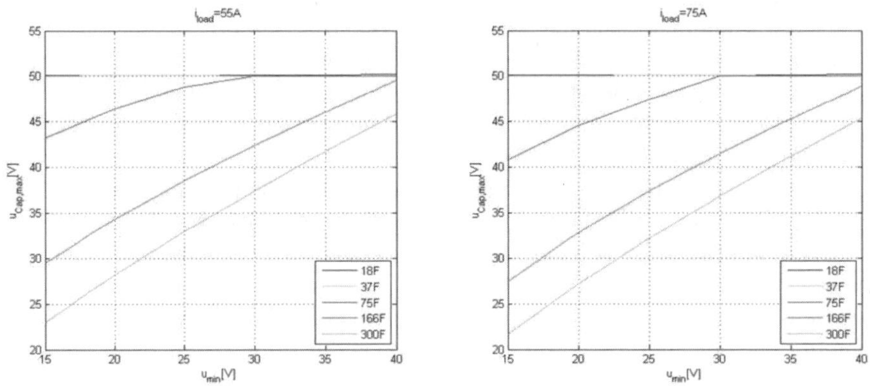

Fig. 10 Maximum voltage U_C reached during a drive cycle versus U_{Cmin} for various C and I_{Load}

Particular scenarios of the simulation have been measured using the measurement set up (Figure 6) as described in Section 4. As an example, Figure 11 presents the measured and the simulated capacitor voltage vs. time and shows a good correlation. The results shown in Figure 9 and 10 allow the calculation of the required installed capacitance kC_{cell}, which determines the capacitor costs, e.g.:

C [F]	U_{Cmin} [V]	U_{Cmax} [V]	U_{Cell} [V]	$k\ C_{Cell}$ [kF] @ 2,7V
18	18	50	2,7	6,2
37	22	50	2,7	12,7
166	32	37	2,7	31,2

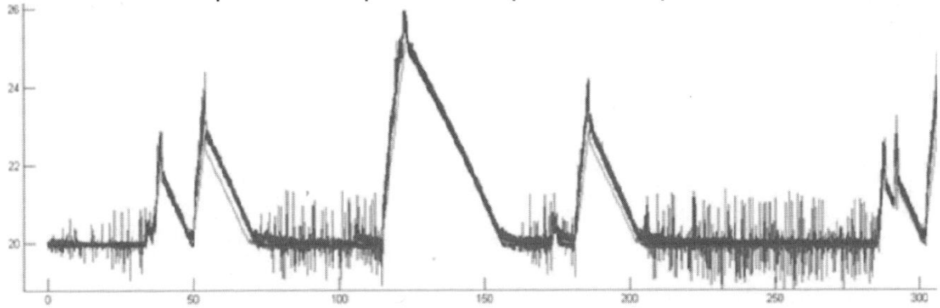

Fig. 11 Simulated and measured capacitor voltage U_C versus time.

8. Conclusion

The proposed architecture provides recuperation as well as a general improvement of the EDS characteristics. However, the components must be carefully dimensioned in order to optimize the balance between cost and performance. A simulation model and a measurement set up are presented to allow both the concept optimization and validation.

References

[1] Richard, D., Dubel, Y.: Valeo StARS Technology: A Competitive Solution for Hybridization", 1-4244-0844-X/07, IEEE 2007

[2] Winkler, J., Esch, S. : "Mikrohybrid ohne Starter Generator mit den Funktionen Schnellheizung und Rekuperation", Conference Baden Baden 2005

[3] Kurz, A.: "Ultracapacitor alternative solution with starter batteries"; Automobil Elektronik, Aug. 2006, p.28

[4] Schneuwly, A., Prummer, M., Auer, J., "Ultra Capacitors: New Energy Storage Concepts in Automobiles", VDI Berichte Nr. 2000, 2007

[5] Neugebauer, St., El-Dwaik, F., Hockgeiger, E., Mattes, W., „Intelligent Contributions for efficient Dynamics of the BMW fleet – the Auto-Start-Stopp Function (ASSF) and Brake-Energy Recovery (BER). VDI Berichte Nr. 2000. 2007

[6] Brabetz, L., "Die Zukunft der Bordnetzarchitektur", Bordnetzarchitekturen, Konsequenzen für das Energie-management, Euroforum, München, April 2002

[7] Miller, J.M., McCleer, J.P., Cohen, M., „Ultracapacitors as Energy Buffers in a Multiple Zone Electrical Distribution System", Maxwell Technologies, White Paper, 2004, http://www.maxwell.com/pdf/uc/white-papers/energy_buffers.pdf

[8] Ploumen,S., Urhahne,J., Spijker,E., Karden,E., Fricke,B.: „Regenerative Braking, Possibilities of Brake Energy Recovery", 3rd Int. Automotive EE Systems, 2007

[9] Rhode-Brandenburge, K.: "Verfahren zur einfachen und sicheren Abschätzung von Kraftstoffverbrauchs-potenzialen", Haus der Technik, 1996
[10] Brabetz,L., Ayeb,M., Tellmann, D.: Efficient vehicle power supply by adaptive energy, charge and heat management of an alternator - super capacitor system, SAE Congress, Detroit, April 2009

HYBRID CONCEPT FOR ENERGY REDISTRIBUTION WITHIN TRANSPORTATION SYSTEMS

L. Britvin[1], B. Schapiro[2]

[1]MADI, Faculty for Engine Design
Moscow, Russia
E-Mail: brt-tgm@mail.ru, Tel.: +7495 7036441

[2]RKM, RotationsKolbenMaschinen GbR
Berlin, Gemany
E-Mail: boris@rkm-schapiro.org, Tel.: +49172 3008947

Abstract The appearance of new, high efficiency volume displacement machines in the low and middle power ranges, such as the Schapiro engine, enables revision and generalization of the hybrid principles. At the fuel station, the basic idea is to store energy from the electrical power grid in batteries and in other forms of stored heat energy, such as the overheated steam of various fluids or salt solutions, for example. A small amount of fossil fuel should also be stored. The primary energy source for transportation should be the stored heat energy. Extracting power from the heat energy store requires a temperature maintenance system for keeping energy extraction at peak efficiency. This temperature maintenance system can be powered by battery, by the chemical energy in a fossil fuel or by solar energy. This concept has advantages even though the overall efficiency of current heat utilization technology is no better (and for all practical purposes not worse) than the efficiency of the transportation technology employed today. A major advantage is that the energy sector is then divided into two almost independent blocs – the producers and the users. Technical and ecological improvement in both blocs can proceed independently. The article contains a methodologically sound comparative table capable of handling both objective and subjective factors in numeric form. Every observer can create a table reflecting his or her personal views.

1. The Energy Unification for Mobility Concept (EU M Concept)

That the EU in the abbreviation for Energy Unification for Mobility coincides with the EU abbreviation for the European Union is not only a gag but rather a meaningful congruence: The EU M concept is intended to be very much in keeping with the spirit of EU environmental policies.

J. Wellnitz et al. (eds.), *Sustainable Automotive Technologies 2010*,
DOI 10.1007/978-3-642-10798-6_32, © Springer-Verlag Berlin Heidelberg 2010

The goal of the EU M concept is the substantially more efficient and environmentally friendly use of energy sources. To this end, we recommend basing the transportation energy market on electricity and restructuring the transportation energy market such that the interface between energy production and energy utilization is unified to a significantly greater extent.

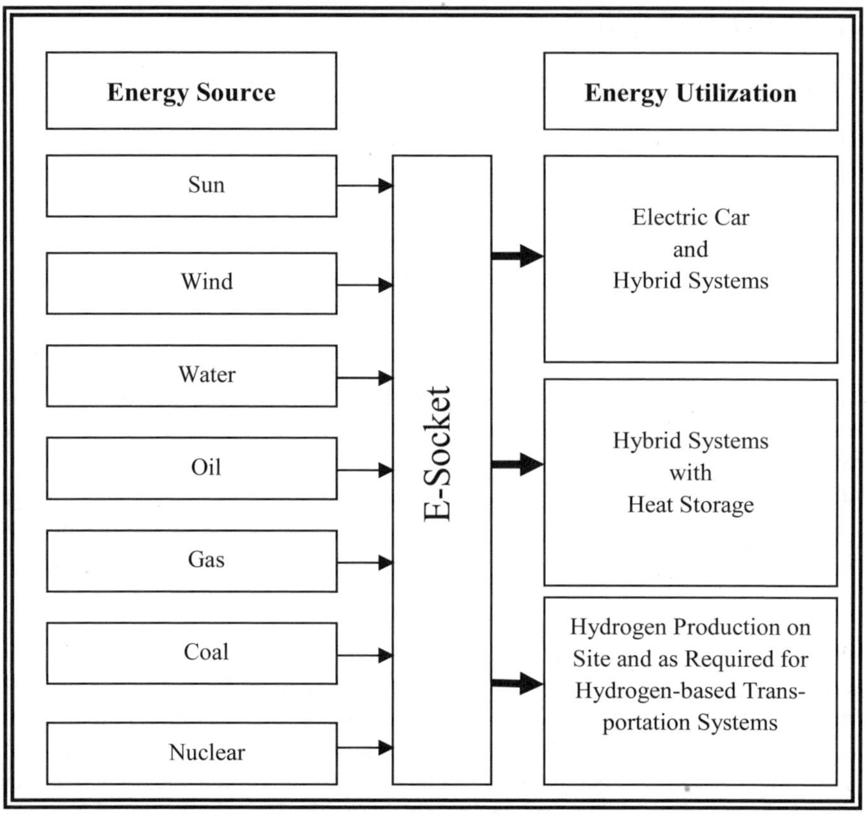

Fig. 1 Recommended Energy Flows for the EU M Concept.

Implementation of the EU M concept offers the following advantages.

Advantages:

- The energy efficiency of fossil fuels will increase by 10 – 15%
- The cleaning of waste gases, as well as heat and resource reclamation, can be made significantly more efficient and cost effective in an industrial production plant. Thus, even though the same amount of fuel may be burned, the environmental footprint can be reduced measurably.
- Transportation costs for fuel delivery will be reduced
- The infrastructure for energy distribution is already in place everywhere

- Future changes in the energy production sector can be implemented without resulting in undesirable costs for required technical changes in the energy utilization sector. The energy production and energy utilization sectors are uncoupled, resulting in greater utility for consumers and the economy as a whole.

To implement the EU M concept requires making changes on both sides of the e-socket. These changes should proceed in parallel, dynamically balanced. Currently, the energy production sector is better prepared for implementing such changes than is the energy utilization sector. The energy utilization sector does not command a diverse fleet with dynamic capabilities to use energy from the e-socket efficiently.

Research and development is already underway and pioneer products already available in the electric car and electric hybrid system sector. Transportation options based on hydrogen are at the stage of prototype development and the testing of small series. However, the potential of these developments meets but a fraction of the transportation demand, especially in the sectors of freight transfer, construction machinery of various kinds and agriculture.

Completely absent is the development of hybrid systems with heat storage for the ground transportation sector. In our opinion, hybrid systems with heat storage are eminently suited for this sector.

2. Transportation Systems with Heat Storage

A wide variety of ways for incorporating heat storage into the optimization of ground transportation systems can be imagined.

The combustion of one liter of diesel fuel in an industrial generating facility can provide 10 – 15% more effective power for the user than combustion of that same liter in a machine for personal mobility.

There will always be a need for on-site combustion of fossil fuels. But that share of the energy mix can be reduced substantially by implementing the EU M concept.

2.1 Heat Storage as a Secondary Energy Source

One possibility for employing heat storage in combination with conventional propulsion aims to increase substantially the efficiency of combustion in the engine. For all practical purposes, the engine should operate continually at peak efficiency. Load increases such as acceleration or driving up hill should be assumed by an ancillary steam engine. The steam engine would be powered by steam from the heat storage facility. In order to maintain the super-heated steam in the heat storage facility at constant operating conditions, heat captured from the main engine by a heat exchanger could be used; electrical power from a battery, from solar panels on the roof of the truck during the day, from electrical outlets at gas stations, chemical heating in case of need, etc., could also be employed.

Steam with its water exhaust is not the only medium that could be employed. Other heat storage and transfer media such as freons, saline solutions, etc., could be used in closed systems.

Reclamation of exhaust heat with an exit temperature of 600 °C for maintaining optimal operating conditions in the heat storage facility could raise the effective exploitation efficiency of the chemical energy stored in the diesel fuel up to 80%. Exhaust pollutants would be reduced correspondingly.

The use of heat storage facilities (heat accumulators) and ingenious steam generation processes coupled with advanced steam engines with thermodynamic efficiencies of approximately 40% should produce the following advantages:

- Considerable increase in the operating efficiency of the primary diesel engine as well as reduction of exhaust pollutants resulting from the engine operating continually at its optimal stationary efficiency while the ancillary steam drive takes over all acceleration and other short-term load increases
- The internal combustion engine can remain turned off from minutes to hours, with propulsion provided solely by the steam drive, should environmental conditions require completely exhaust-free operation such as in danger zones, cellars or underground garages or other enclosed spaces, congested areas, etc. The same holds true for acceleration from a stop (such as at a traffic light) or other increased load situations during which diesel motor efficiency is lowest and exhaust toxicity highest.
- Heat accumulators can be recharged with waste heat from the primary engine, as well as with solar energy, power from the electrical net or energy from other sources.

The combination of diesel engine plus steam engine with heat storage facility is particularly expedient for use with heavy-duty trucks for which efficiency optimization and reduction of exhaust toxicity is particularly important, not only for the environment for also for the energy economy.

2.2 Heat Storage as Primary Energy Source

3.4 tons of water at 375 °C and 220 atm, plus 25 liters of diesel fuel for maintaining optimal power output, can power a fully loaded 40-ton truck for 10.5 hours at 100 kilometers/hour under optimal conditions (level road and a 40% efficient steam engine) for a total range of 1000 kilometers.

In this scenario, 2.5 to 3 liters of diesel fuel per 100 kilometers travelled are required to maintain optimal power output, compared to 35 to 45 liters per 100 kilometers required to travel with the diesel engine alone.

As an example, let's take a 40-ton truck, travelling a distance of 1,000 kilometers fully loaded. With conventional diesel technology, this 1,000 kilometer road trip will require approximately 425 liters of diesel fuel.

With the EU M concept, the same 40-ton truck would use only 25 liters of diesel fuel for the same stretch. Of course, one must also include the approximately 385 liters of diesel fuel required if the electricity used to heat up the heat storage system initially is generated industrially using diesel fuel only.

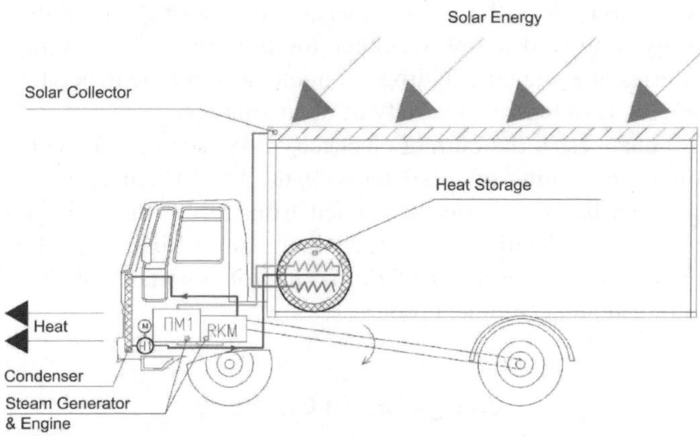

Fig. 2 Possible configuration of steam-driven vehicle with heat storage system and steam generator.

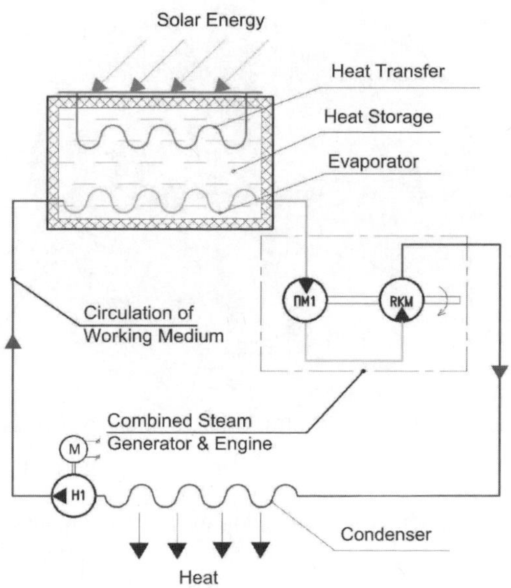

Fig. 3 Functional schematic of possible configuration.

Saving 15 liters of diesel fuel per 1,000 kilometers with the EU M concept may not seem like a major improvement. Although that doesn´t seem like much, it adds up quickly. In Germany, for example, with roughly 200,000 registered 40-ton tractor-trailers driving an average of 40,000 kilometers per year, that adds up to 120,000,000 liters of diesel fuel per year. At current prices, that is 132 million Euros needlessly burned, assuming that the electricity used to preheat the heat storage facility was generated using diesel only. A different electricity mix would result in even greater waste.

The greatest strength of the EU M Energy Unification for Mobility concept is that the energy delivered to the e-socket for preheating heat storage facilities, charging batteries or producing hydrogen on-site need not be generated by burning fossil fuels but can come from a variety of other sources.

Long-term changes in the current electricity mix (see Fig. 3) will certainly include reducing the burning of fossil fuels. If the EU M concept is implemented, these changes can be completely decoupled from further technological developments in the energy utilization sector (the right side of Figure 1). This could save the world economy many billions of Euros in avoidable costs without detracting from technological and ecological progress.

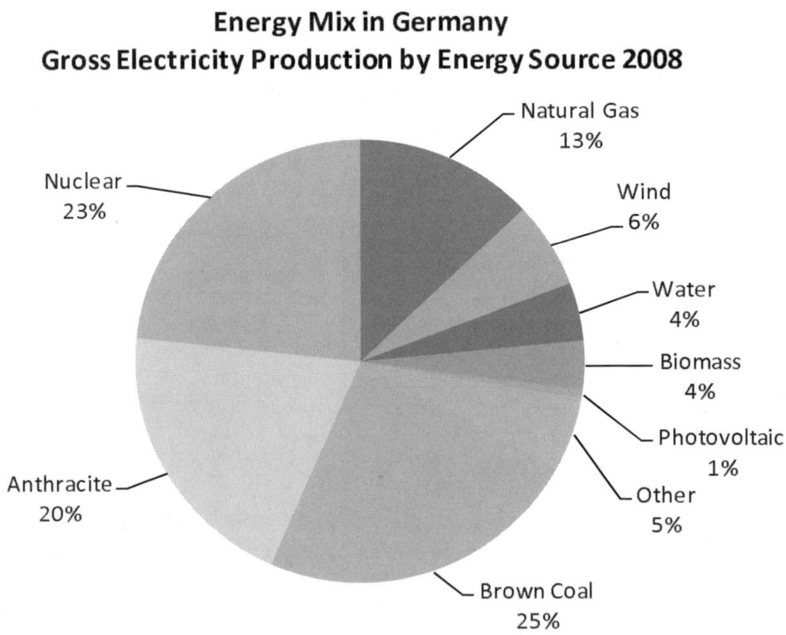

Fig. 4 German electricity mix from http://de.wikipedia.org/wiki/Stromkennzeichnung.

2.3 Heat Storage in Combination with a Porous Burner

The porous burner technology developed at the University of Erlangen enables high-pressure steam injection for the newest generation of steam engines at pressures comparable to common rail systems; porous burner fuel economy is similar to that of a 90 horsepower TDI motor (Emo-Auto 2009).

A scientific prototype based on the Škoda Fabia proves the great potential of porous burning, at least for steam generation. Steam temperatures up to 900 °C are created. The burner/steam generator system delivers steam with pressure, temperature and volume adjusted for load and rpm differences to meet power requirements. With the experimental facility called ZEE03, steam is injected at up to 500 atm – injector opening and closing times are under one millisecond. The injector's open time controls steam volume and, thereby, engine power output; the expanding steam drives the piston.

3. Conclusion

Traffic causes the lion's share of emissions (see http://www.stadtklima-stuttgart.de/index.php?klima_klimaatlas_6_emissionen - German only) harmful to man and climate in so many ways. In the energy balance (well-to-wheel) with the current power generation mix, electric drives are more efficient than internal combustion engines and can therefore contribute to decreasing CO_2 emissions. However, substantial advantages for our climate can be achieved only when electricity is generated with sources other than fossil fuels.

Achieving this goal with the least economic stress requires the maximum possible structural decoupling of the energy production and energy utilization sectors. Only then can development in both sectors proceed in parallel without imposing costly, reciprocally necessary decisions. This is a crucial element of the EU M concept.

The energy production sector's development strategy is not at issue here whereas that of the energy utilization sector certainly is. What we really need is a comprehensive study of the cost-benefit relationships, differentiated and weighted according to usage, emissions, type of engine, vehicle weight, mission, kilometers driven, substitutability of new technologies, operator adoption, user acceptance, political and social temper, etc. We could not conduct such a study on our own as it requires serious financing.

Instead, we have attempted an abstract comparison of technology development strategies for the energy utilization sector. By setting artificial values and multipliers (weighting factors) for the various strategies, we attempt to evaluate strategies difficult to compare (see the table in Figure 5).

The values reflect the current state of technology and the opinions of various experts in the respective areas. The choice of weighting factors expresses our subjective opinion which, of course, is considerably influenced by our own view of the world.

The comparative methodology is, however, quite transparent. Thus, anyone can generate his or her own comparative table as a decision aid simply by inserting personal priorities in the weighting factor hierarchy. Moreover, with the help of sociological data collection, a profile of various population groups' preferences could be generated to research the range of acceptance and estimate possible future buyer acceptance.

Based on our comparison, we conclude unequivocally that, if diesel and natural gas-based transportation technology is rejected for the greater good, then steam-diesel-electric-fossil hybrid technology has an absolute advantage (74 total points, 0.822 normalized value), with electric-fossil hybrid technology coming in second (65 total points, 0.722 normalized value).

Characteristic	Characteristic Weighting Factor	Ideal Technology	Conventional		Electric		Hybrid	
			Fossil-fluid Diesel	Fossil-gasf	Battery	Fuel Cell	Electric & Fossil	Steam & Accessory EU M Concept
Overall Efficiency	4	5	5	3	3	3	4	4
Drive Efficiency	4	5	4	3	5	4	4	3
Global Environmental Load	3	5	4	5	2	3	3	5
Local Environmental Load	2	5	3	4	5	5	4	4
Range	2	5	5	3	1	3	4	5
Cost per Kilometer Driver	2	5	4	5	2	1	3	4
Vehicle Cost	1	5	5	5	1	2	2	5
Absolute Weighted Total		90	77	68	55	57	65	74
Normalized Weighted Total*		1	0.856	0.756	0.611	0.633	0.722	0.822

* Rounded

Fig. 5 Qualitative estimates on a scale of 1 (disadvantageous) to 5 (very advantageous) weighted by factors from 1 (not very important) to 4 (very important) based on our subjective perceptions.

We expect the efficiency-increasing utilization of Schapiro machines, rotating piston machines with a jumping instantaneous axis and other geometries (www.rkm-schapiro.org), to further enhance the advantages of both hybrid technologies.

Other observers can construct their own comparative tables as decision aids by expressing their own priorities in the hierarchy of weighting factors. Although the ratings of various observers will become difficult to compare should a different absolute value for the ideal technology be calculated, normalizing the value ensure comparability.

The utility of normalizing weighted total value totals to 1 (one) is that the ratings of various observers coming to different total values can be correctly compared to one another in a uniform fashion. Such comparisons will be meaningful and could be a useful basis for constructive discussions.

Acknowledgments The authors would like to thank Karl Sittler for his help in preparing the English text..

References

Emo-Auto (2009). www.emo-auto.de. Accessed on 9th December 2009. Available at http://www.emo-auto.de/Technik/Antrieb/Antr_3_4_3/antr_3_4_3.html#Dampfmotor.

Schapiro B. (2008), The RPM Rotary Piston Machines. In: Plath, P. J. und Hass, E.-Ch. (Edrs), Vernetzte Wissenschaften, Logos Verlag Berlin 2008.

Schapiro B. and Terlitsky L. (2008), RKMs: New Class of Machines. In: Subic A., Leary M. and Wellnitz J. (Edrs), Meeting the Challenges to Sustainable Mobility, Proceedings ICSAT 2008, International Conference of Sustainable Technologies, November 2008, Melbourne, Australia.

RKM GbR Website: www.rkm-schapiro.org.

A SYSTEM DYNAMICS APPROACH TO CALCULATING THE ENERGY CONSUMED IN CAR COMPONENT PRODUCTION

P. Stasinopoulos, P. Compston, H. M. Jones

School of Engineering, College of Engineering and Computer Science,
The Australian National University, Canberra, ACT 0200, Australia,
E-mail: u4577906@anu.edu.au, paul.compston@anu.edu.au, haley.jones@anu.edu.au

Abstract Standard Life Cycle Assessment calculation procedures may be limited in their ability to reflect the reality of a product system because, typically values that are variable are assumed to be constant or to follow a fixed trajectory. This limitation can be overcome by using system dynamics, which is an approach in which the calculation tools allow the value of variables to change as the characteristics of the systems change. This paper presents an example that uses system dynamics approach to compare the energy consumed in producing car components from aluminium and steel.

1. Introduction

The ISO Life Cycle Assessment (LCA) technique (Standards Australia and Standards New Zealand 1998) is used to evaluate various environmental impacts of a product system. This paper focuses on Phase 2 of LCA, which involves data collection and calculation procedures to quantify relevant inputs and outputs of a product system.

The literature on standard LCA case studies indicates a strong tendency towards calculation procedures for Phase 2 in which the values that are variable are assumed to either be constant or follow a fixed trajectory. However, these variables are rarely fixed, especially for new products. Rather, they vary over time under the influence of the complex feedback effects arising from the interactions with other components in the wider system. Accounting for this influence requires the use of an alternative approach in Phase 2. One such approach is system dynamics (Sterman 2000).

Using system dynamics, particularly the stock-and-flow modelling tool, can provide more realistic estimates of a product's environmental impact than using standard procedures. Accurate estimates are particularly important for products that have a relatively long life cycle, a high production volume and a large environmental impact. One such product is the passenger car, which contributes consi-

J. Wellnitz et al. (eds.), *Sustainable Automotive Technologies 2010*,
DOI 10.1007/978-3-642-10798-6_33, © Springer-Verlag Berlin Heidelberg 2010

derably to energy consumption and air pollution. This paper will demonstrate the use of system dynamics in Phase 2 of LCA with an example that compares the energy consumed in producing car components from aluminium and steel.

2. Literature Review

All LCA case studies reviewed use standard Phase 2 procedures for which the variables are artificially fixed. In fact, the Australian/New Zealand Standard for LCA, itself, includes an example that uses fixed inputs and outputs (Standards Australia and Standards New Zealand 2001). Most of the LCA studies on car materials show that cars made with aluminium components consume less energy over their life cycles than cars made with the equivalent steel components, with the higher energy investment of aluminium production being paid off during operation and then largely recovered at end-of-life. In these studies, there are differences in the scopes of the product systems and in assumptions about variables. Most studies focused on a single car (product-based), with most assuming that variables are constant (e.g. Puri *et al.* 2009) and some assuming that variables follow a fixed trajectory (e.g. Ungureanu *et al.* 2007). A relatively small number of studies focused on the whole car fleet (fleet-based), with most assuming that market penetration rates of aluminium components are constant (e.g. Das 2000) and a few assuming that market penetration follow a fixed trajectory (Field *et al.* 2000).

Replacing the standard Phase 2 procedures with a system dynamics evaluation is consistent with the LCA standard, which does not prescribe the calculation procedures (Standards Australia and Standards New Zealand 1998). A system dynamics evaluation includes the feedback interactions between the product and its wider system that comes into play after the product has been introduced to the market. Incorporating these interactions into the evaluation affect the results in two ways. First, considering the feedback effects may be more accurate than artificially fixing the variables. The feedback effects are key characteristics of the real system, often dominate the system's behaviour, and are often the source of unexpected behaviour (Ashford 2001). Second, considering the wider system is more informative than limiting the scope to the product only. A wider spatial and temporal scope enables a measure of the total impact of a new product in the market (Das, 2000; Field *et al.* 2000).

3. Example: Energy consumed in producing car body-in-whites

Consider the substitution of aluminium 6061-T6 for mild steel as the primary material in a car's body-in-white (BIW), which is the load-carrying welded frame to which other moving components are attached. In this example, the scope is limited to the production phase of BIWs, which represents the energy investment tied up in cars, and the measure of environmental impact used is the energy consumed in producing the BIWs.

Cars have been developed with aluminium BIWs of mass 20-45% less than the equivalent steel BIWs (Kelkar *et al.* 2001). In this example, the aluminium BIW has a mass of 300 kg and the steel BIW has a mass of 430 kg. Aluminium requires more energy to produce from virgin material (190 MJ/kg) than steel (34 MJ/kg) (Puri *et al.* 2009). However, the energy required to produce aluminium from recycled materials (9.5 MJ/kg) is comparable to that of steel (8.8 MJ/kg) (Bureau of International Recycling 2009).

3.1 The standard LCA approach

Most standard LCA studies calculate the energy consumed in producing a single aluminium BIW, usually for a single car, using a fixed mixture of 0-15% recycled and 85-100% virgin content. The mixture is assumed to be 10% recycled and 90% virgin aluminium, which, given the above values of energy consumption and mass of the BIW, indicates that the energy consumed in producing a single aluminium BIW is 51.6 GJ.

3.2 The system dynamics approach

The energy consumed in producing the BIWs in an average car depends on the number of cars with recycled aluminium BIWs (*RecAl Cars*), virgin aluminium BIWs (*VirAl Cars*) and steel BIWs (*St Cars*) at the time. These variables are estimated using a stock-and-flow model. The model that estimates the number of *RecAl Cars*, *VirAl Cars* and *St Cars* is shown in Figure 1, where rectangles represent stocks of cars and materials, pipes represent the flow of stocks, valves represent the rates of flows, clouds represent unlimited sources or sinks of stocks, circles represent other variables, arrows represent influence of one variable on another, and computation is embedded in valves and circles.

The model makes the following simplifying assumptions: aluminium BIWs are new to the car industry and their diffusion is instantaneous, meaning that all current BIWs are made from steel and that all new BIWs will be made from aluminium; all producers favour the use of recycled aluminium over virgin aluminium, when available; the stocks of virgin aluminium and steel are unlimited; and the only source of recycled aluminium is end-of-life vehicles. The model is comprised of the following rows of stocks and flows:

- Row 1 shows the flow of *St Cars* out of the production phase, into the operation phase and into the end-of-life (*EoL*) phase. The flows are regulated by the *Growth Rate of St Cars,* expressed as a fraction of the current *Total Number of Cars*, and *Attrition Rate of All Cars*, expressed as a fraction of the current *Number of St Cars*.
- Row 2 and 3, in similar fashion to row 1, show the flows of *RecAl Cars* and *VirAl Cars*. The production of *RecAl Cars* is prioritised over *VirAl Cars* by applying a different growth rate to each. The resulting behaviour is that the maximum available *Number of RecAl Cars per Year* are produced and then, since the stock of virgin aluminium is unlimited, the shortfall is met by *Number of VirAl Cars per Year*.
- Row 4 shows the flow of aluminium from *Number of EoL Al Cars* into *Amount of RecAl* and then into *Number of New RecAl Cars per Year*. The resulting behaviour is that the *Amount of RecAl* limits the *Number of new RecAl Cars per Year* produced. The loop also factors in the *Recycle Rate of Al*, since it may not be technically or economically feasible to recycle all of the aluminium.

The model produces the results in Figure 2, given: *Growth Rate of Fleet* = *Attrition Rate of Fleet* = 0.04/year, which makes the *Total Number of Cars* constant for demonstration purposes; *Growth Rate of St Cars* = 0, since all new cars are *RecAl Cars* or *VirAl Cars*; *Number of St Cars* = 12 million (initially); *Recycle Rate of Al* = 0.9.

The results in Figure 2, along with the data presented above and assuming that steel BIWs are comprised of 25% recycled and 75% virgin content, provide the energy consumption of producing BIWs in operation, as shown in Figure 3.

The model indicates that the time taken for *RecAl Cars* to make up 10% of all aluminium cars in the fleet is 8 years. At this point, *St Cars* make up 73% of all cars, *Total Energy in Cars* is 2.7×10^8 GJ and the energy consumed in producing the average BIW in the fleet is 22.9 GJ.

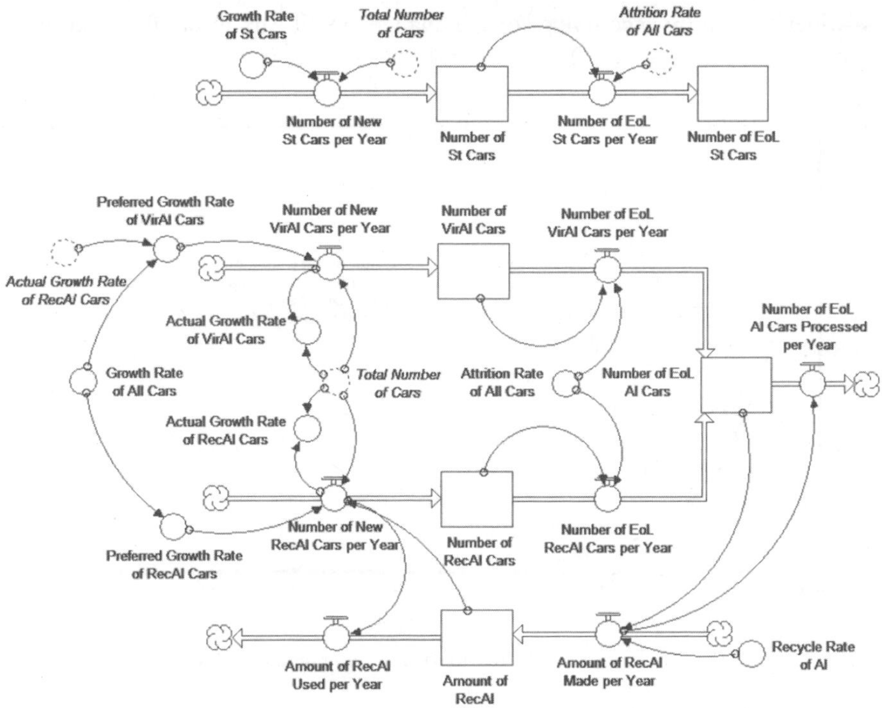

Fig.1 A simplified stock-and-flow model of the life cycle of steel and aluminium cars.

4 Discussion

The product-based standard LCA calculation indicates that the energy consumed in producing a single aluminium BIW with 10% recycled content is 51.6 GJ. This evaluation is of the steady state that assumes that all cars in the fleet have BIWs with a fixed mixture or recycled and virgin content. Extrapolating this calculation across the fleet of 12 million cars gives a *Total Energy in Cars* of 6.2×10^8 GJ.

By comparison, the fleet-based stock-and-flow model indicates that the energy consumed in producing the average BIW at the point when 10% of the aluminium cars have recycled BIWs is 22.9 GJ. This evaluation also indicates that, after the introduction of *Al Cars* to the market, there is a dominant transient period in which the total energy consumed in producing the BIWs in the fleet peaks at 3.4×10^8 GJ before reaching steady state at 3.4×10^7 GJ.

A comparison of the two calculation approaches highlights two main advantages of the system dynamics approach. First, it is more accurate because it considers the feedback effects by accounting for the accumulation of a useful stock of 'low-energy' recycled aluminium. Second, it is more informative because it considers the wider system by accounting for the changes in all cars in the fleet, not just a single car.

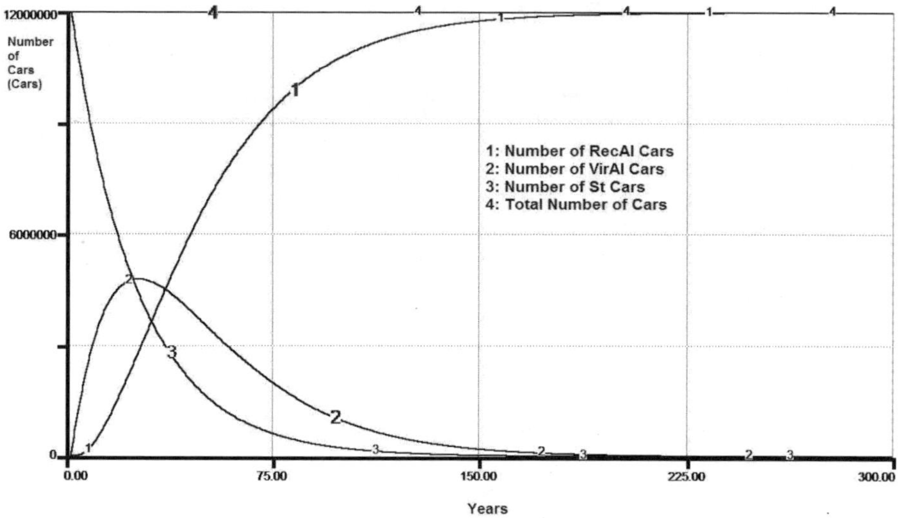

Fig. 2 Changes in the numbers of cars after introducing aluminium BIWs.

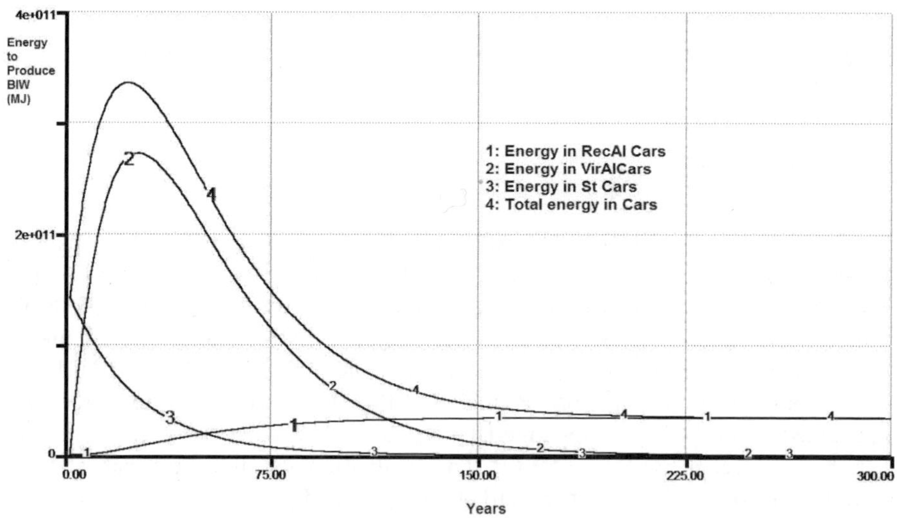

Fig. 3 Changes in the energy consumed in producing BIWs after introducing aluminium BIWs.

5 Conclusion

The limitations of the standard calculation procedures in Phase 2 of LCA can be overcome by using a system dynamics approach. This paper presents an example that uses a system dynamics approach to compare the energy consumed in producing car components from aluminium and steel. The example highlights that using this approach can provide more accurate and more informative results than using standard calculation procedures in which variables are artificially fixed.

Acknowledgments The authors would like to thank Dr Barry Newell for his insights and review of this paper; and the Australian National University and the AutoCRC for their financial and administrative support into the development of this paper.

References

Ashford A C (2001) *Unexpected behaviors in higher-order positive feedback loops*. Massachusetts Institute of Technology, USA (online). Available: http://sysdyn.clexchange.org/sdep/Roadmaps/RM7/D-4455-2.pdf. (Accessed: 24 September, 2009).

Bureau of International Recycling (2009) *About recycling*. BIR, Belgium (online). Available: http://www.bir.org/aboutrecycling/index.asp. (Accessed: 13 September, 2009).

Das S (2000) The life-cycle impacts of aluminum body-in-white automotive material, *Journal of the Minerals, Metals and Materials Society*, 52/8, 41–44.

Field F, Kirchain R and Clark J (2000) Life cycle assessment and temporal distributions of emissions: developing a fleet-based analysis, *Journal of Industrial Ecology*, 4/2, 71-91.

Kelkar A, Roth R and Clark J (2001) Automobile bodies: Can aluminum be an economical alternative to steel?, *Journal of the Minerals, Metals and Materials Society*, 53/8, 28-32.

Puri P, Compston P and Pantano V (2009) Life cycle assessment of Australian automotive door skins, *International Journal of Life Cycle Assessment*, 14/5, 420-428.

Standards Australia and Standards New Zealand (1998) *Australian/New Zealand Standard: Environmental management – Life cycle assessment – Principles and framework*, AS/NZS ISO 14040:1998, Standards Australia and Standards New Zealand.

Standards Australia and Standards New Zealand (2001) *Australian/New Zealand Standard: Environmental management – Life cycle assessment – Life cycle interpretation*, AS/NZS ISO 14043:2001, Standards Australia and Standards New Zealand.

Sterman J.D. (2000) *Business dynamics: Systems thinking and modeling for a complex world*, McGraw Hill.

Ungureanu C A, Das S and Jawahir I S (2007) Life-cycle cost analysis: Aluminum versus steel in passenger cars. In Das S K and Yin W (Eds.) *Aluminium alloys for transportation packaging aerospace and other applications*. TMS, USA.

Part 8
Vehicle and public safety

VEHICLE AND PUBLIC SAFETY THROUGH DRIVER ASSISTANCE APPLICATIONS

M. Simic

RMIT University, Bundoora, Victoria 3083, Australia
Phone: +61 3 9925 6223, E-mail: milan.simic@rmit.edu.au

Abstract This paper presents a Driver Assistance Application research and development. With the integration of communications technologies and mechatronics engineering, vehicles are becoming truly autonomous, intelligent transport systems. A short review of the variety of emerging applications is given. Our approach is that the advantages of new technologies should be available to all drivers and include any vehicle already on the road. It is possible thanks to the amalgamation of computing and communications technologies expressed in mobile communications devices and PDAs. This is the hardware platform that we utilised while building the system which can be used for the further development of new algorithms and applications.

Applications overview and project objective

The need to increase vehicle and public safety is motivating development of new automotive technologies and research in other supporting systems. They are known as safety systems, passive and active, and driver assistance systems. The short list of emerging applications is given here: Vehicle to Vehicle communications (V2V), or Vehicle ad Hoc Networking (VANET) (Reichardt et al. 2002), Navigation systems, with the use of General Positioning System (GPS), or Traffic Message Channel (TMC); Collision avoidance systems, Night Vision, Adaptive light control; Blind spot warning systems, Traffic sign recognition; Cruse control, Driver drowsiness detection; Lane Detection (D'Cruz et al. 2007), Parking assistance; Novel Human Machine Interfacing (HMI) Solutions (Doshi et al. 2009); Electronics steering control and many others.

For example, one of the new solutions for HMI is dynamic active display (DAD). This unique interface presents safety-critical visual icons to the driver without obstructing other sources of information, i.e. currently used presentation media. Driver assistance applications often use vision and image processing to monitor the driver behaviour, or predict other drivers' actions (Baehring et al. 2005). Visual detection is a complex task. It will play central role in the vehicles' full awareness of the traffic conditions (Fritsch et al. 2008). A comprehensive re-

J. Wellnitz et al. (eds.), *Sustainable Automotive Technologies 2010*,
DOI 10.1007/978-3-642-10798-6_34,

view of different object recognition methods for the application in driver assistance systems is given in (Laika et al. 2007).

Vehicle can be seen as a mobile robot. Moral, ethical and primarily safety question is, how much of the vehicle control should be given to the computer system. An adaptive authority adjusting strategy, that increases the coordination and cooperation between driver and computer system, was considered in (Xiaoming et al. 2009).

Path planning algorithms, from the robotics theory, are often applicable to vehicles parking scenarios. Kinematics of the parking vehicle problem can be defined by formulations already used in robotics. A closed loop approach in the solution of the semi-autonomous nonholonomic vehicle parking problem is shown in (Oetiker et al. 2009). Parking is performed by the use of the arrangement of calculated vector fields, of feasible nonholonomic paths, by which the vehicle is controlled. A nonholonomic system is fully controllable, if it is small-time controllable. System depends on the sensors array feedback information to avoid collisions. Our objective in this project was to design such sensor area and control system, which could be used for testing various scenarios of motion control, i.e. parking assistance algorithms.

1.2. Control systems

A block diagram of a driver assistance application, is given in Figure 1. Sensor system may include different types of sensors mounted on the vehicle. We could have radar, ultrasonic, infrared, laser or other type of transducers. We already have large number of sensors and actuators used in the vehicle engine and motion control, not explicitly shown on the Figure. They are all integrated in the Vehicle Control System. This system also includes number of microcontrollers and number of busses: Controller Area Network (CAN), FlexRay, Media-Oriented Transport (MOST) and Ethernet.

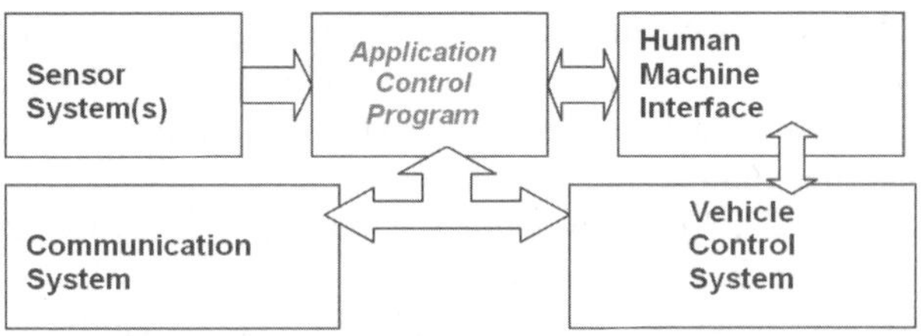

Fig 1. General model of a driver assistance application.

RMIT University research and development

At RMIT University, School of Aerospace, Mechanical and Manufacturing Engineering (SAMME) research in active safety is conducted through the number of projects performed by different research groups. Results presented here are outcomes of the research activities in the International Masters study of automotive electronics. Projects like: Parking Assistance, Blind Spot Warning (Simic 2009), Regenerative Breaking and other, were successfully finalised during 2009.

2.1. Project task kinematic model

In order to define kinematic model of the parking vehicle let us assume that the ***V*** is the subset of the workspace ***W***, covered by the vehicle as shown in Figure 2. We further assume that, the front wheels ***F*** are coupled by the Ackermann angles, so that the steering could be represented by the single angle *Φ(t)*. We also assume low vehicle speed and acceleration. This is an example of nonholonomic system with rolling without slipping constraints (Latombe 1991; Spong et al. 2006). They are defined by setting the sideways velocity of the front and rear wheels to zero:

$$\dot{\boldsymbol{x}}(\boldsymbol{t})\sin\boldsymbol{\theta}(\boldsymbol{t}) - \dot{\boldsymbol{y}}(\boldsymbol{t})\cos\boldsymbol{\theta}(\boldsymbol{t}) = 0$$

$$\dot{\boldsymbol{x}}(\boldsymbol{t})\sin(\boldsymbol{\theta}(\boldsymbol{t}) + \boldsymbol{\phi}(\boldsymbol{t})) - \dot{\boldsymbol{y}}(\boldsymbol{t})\cos(\boldsymbol{\theta}(\boldsymbol{t}) + \boldsymbol{\phi}(\boldsymbol{t})) - \boldsymbol{d}\,\dot{\boldsymbol{\theta}}(\boldsymbol{t})\cos\boldsymbol{\phi}(\boldsymbol{t}) = 0$$

Kinematic model of the parking vehicle, as shown in the Figure 2 is:

$$\dot{x}(t) = v(t)\cos\theta(t)$$

$$\dot{y}(t) = v(t)\sin\theta(t)$$

$$\dot{\theta}(t) = v(t)\frac{\tan\phi(t)}{d} = v(t)k_1(t)$$

$$\dot{k}_1(t) = k_2(t); \quad \dot{k}_2(t) = k_3(t), \text{where } |\mathrm{k}_1(t)| \le c_1, |\mathrm{k}_2(t)| \le c_2, and\ |\mathrm{k}_3(t)| \le c_3$$

Comfort state $\boldsymbol{k_2}$ and the acceleration state $\boldsymbol{k_3}$ are defined as given in the kinematic model equations set. Mechanical limitations of the steering angle are expressed as $|\Phi(\boldsymbol{t})| \le \boldsymbol{c}_1 = \Phi_{\max} < \frac{\pi}{2}$

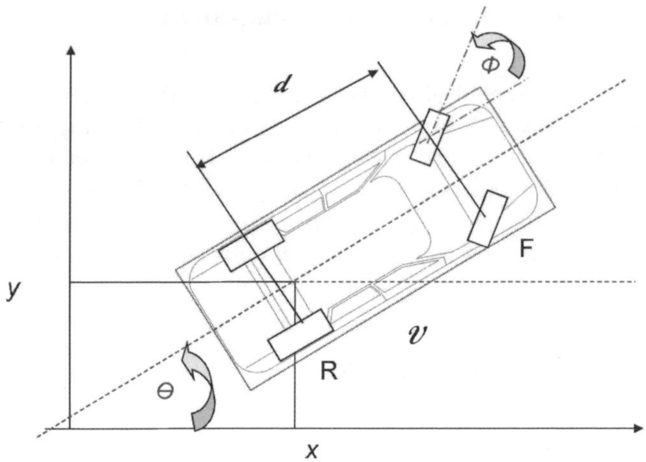

Fig. 2 The kinematic model of a parking vehicle.

Constants c_2 and c_3 refers to the other limits of the actuator subsystem. This is a generic description of the steering system and does not determine its exact dynamics. We will now concentrate to the practical design of the system that could be used for the testing of various path control algorithms.

2.2. Design issues

Few new vehicles are already equipped with Parking Assistance. The level of sophistication and automation depends on the cost, vehicle type and transmission system design. Vehicles with manual transmission require driver input during the process execution. If application control program need to take full motion control of the vehicle, it has to use Vehicle Control System (VCS) resources as shown in Figure 1. The application program can be just one of the tasks performed on VCS, or could run from the independent system platform. In that case it has to communicate with vehicle system via CAN bus interface. There is a simple scenario that does not involve the use of CAN and system resources directly, if HMI is employed. Motion control instructions are given to the driver using visual or audio communication. We have accepted this approach and our efforts were concentrated on input / output system definition, algorithm and program development.

As an input device ultrasonic range finder SRF02 was selected. It is a combined speaker/receiver that can detect objects in the range of 15cm to 6m, with resolution of 3-4cm. Typical frequency used is 40 kHz and it is produced by a piezoelectric actuator. Measurement is conducted by sending and receiving short signal pulses, as shown in Figure 3. A small number of signal periods are sent and the response waited. Time interval from the emitting and the detection is measured, and the distance calculated is $\frac{1}{2}(v_{sound}\Delta t)$.

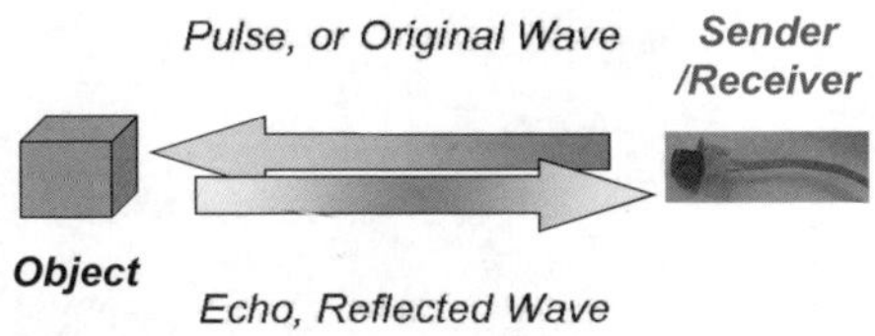

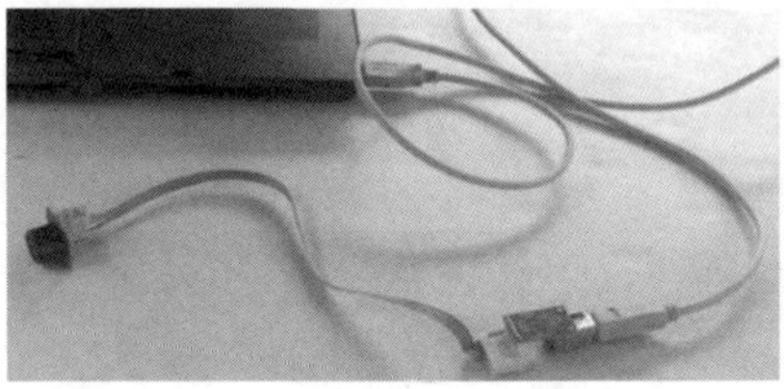

Fig. 3 Distance measurement with ultrasonic sensors.

Sensor can be connected to microcontrollers, or PCs using USB, or serial channel. It performs signal processing and communication with the system controller. Up to 16 ultrasonic sensor modules could be connected to the same bus. Communication is conducted using a set of commands. Each time two communication bytes are issued. The first byte is the module address and second one is the command byte.

2.3 Sensor network layout

The next step was to determine the number of sensors and the sensor network layout. According to the product documentation, the beam divergence angles are ±40° at 34 cm, ±30° at 124 cm and so on, as shown in Figure 4. Following that, we have placed sensors as shown in Figure 5, so that they cover 360°. Central axes of neighbouring transducers intersect at an angle of $\Theta = 45°$. With this space arrangement we have tried to minimise signal interference and to prevent false readings of the returned waves sent by the other transducers. By performing the test measurements we have established that the minimum, reliable distance measured is about 30cm. In Figure 5, sensor addresses are represented in binary. Sensor 0000 faces forward direction, while sensor 0100 looks backwards since its central axis is skewed for 180°. In addition to the spatial and angle discrimination, time separation, i.e. synchronisation, is applied in order to avoid signal interference.

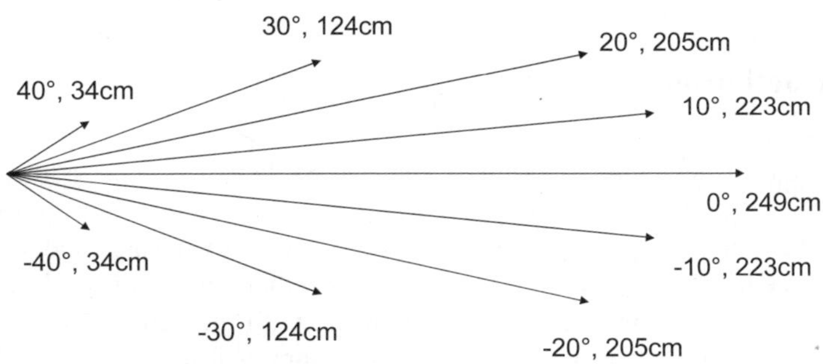

Fig. 4 Beam pattern for SRF02.

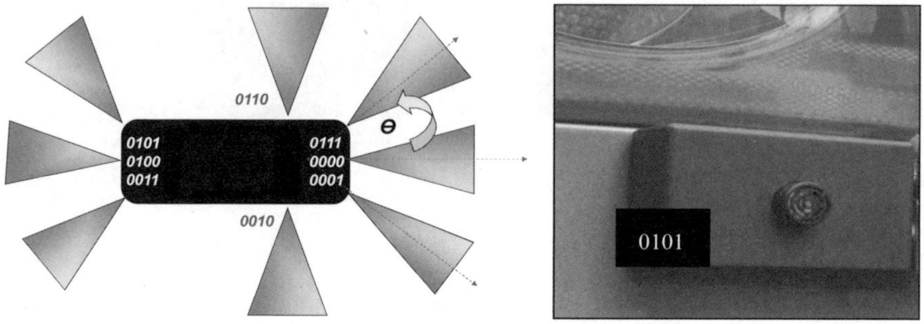

Fig. 5 Ultrasonic sensor area mounted on the car.

2.4. Control program

Control program, designed as a State Machine, is shown in Figure 6. In the initial stage, self test procedures are performed. Next, the search for the available spot is conducted using speed of the car, the length and the maneuvering capabilities. In the third state, program for the reverse parallel parking is performed. It calculates the best steering angle and displays it on the screen. Audio and visual signals are used for the driver feedback.

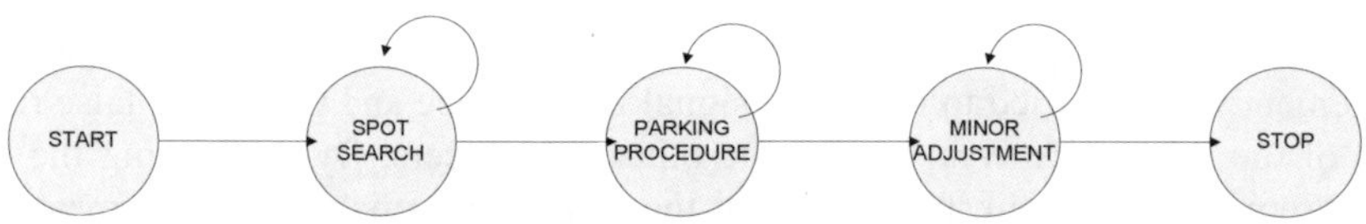

Fig. 6. Reverse Parallel parking control program state diagram.

If minor position adjustments are needed they are performed in the fourth state. In the last state system is deactivated and all outputs are cleared.

Further activities

The objective of designing Parking Assistance System was successfully accomplished. Prototype can now be improved and used for the new algorithms development. There are few pathways ahead. Firstly, the full working model based on mobile device like PDAs should be released. The other pathway would be the next generation of the systems that will communicate with the vehicle on board computer and take full control of the motion. This approach is already developed in robotics and the best solutions are currently being transferred to automotive engineering.

Driver Assistant Applications are a promising and challenging area of research and development, which also include legal, ethical and other discussions and breakthroughs.

References

Baehring, D., Simon, S., Niehsen, W., & Stiller, C. (2005). Detection of close cut-in and overtaking vehicles for driver assistance based on planar parallax. Paper presented at the Intelligent Vehicles Symposium, Proceedings. IEEE.

D'Cruz, C., & Jua Jia, Z. (2007). Lane detection for driver assistance and intelligent vehicle applications. Paper presented at the Communications and In-formation Technologies, 2007. ISCIT. International Symposium on.

Doshi, A., Shinko Yuanhsien, C., & Trivedi, M. M. (2009). A Novel Active Heads-Up Display for Driver Assistance. Systems, Man, and Cybernet-ics, Part B: Cybernetics, IEEE Transactions on, 39(1), 85-93.

Fritsch, J., Michalke, T., Gepperth, A., Bone, S., Waibel, F., Kleinehagenbrock, M., et al. (2008). Towards a human-like vision system for Driver Assis-tance. Paper presented at the Intelligent Vehicles Symposium, IEEE.

Laika, A., & Stechele, W. (2007). A review of different object recognition methods for the application in driver assistance systems. Paper presented at the Image Analysis for Multimedia Interactive Services, 2007. WIAMIS. Eighth International Workshop on.

Latombe, J. C. (1991). Robot Motion Planning. AH Dordrecht, The Netherlands: Kluwer Academic Publishers.

Oetiker, M. B., Baker, G. P., & Guzzella, L. (2009). A Navigation-Field-Based Semi-Autonomous Nonholonomic Vehicle-Parking Assistant. Vehicular Technology, IEEE Transactions on, 58(3), 1106-1118.

Reichardt, D., Miglietta, M., Moretti, L., Morsink, P., & Schulz, W. (2002). Car-TALK 2000: safe and comfortable driving based upon inter-vehicle-communication. Paper presented at the Intelligent Vehicle Symposium, IEEE.

Simic, M. (2009, April). Blind Spot Warning Systems Solutions. Paper presented at the 22nd JUMV International Automotive Conference, Belgrade, Serbia.

Spong, M. W., Hutchinson, S., & Vidyasagar, M. (2006). Robot modeling and control. Hoboken: John Wiley & Sons, Inc.

Xiaoming, Z., Tao, M., Rujing, W., & Xuemin, M. (2009). A kind of cognitive driver assistance system with adaptive authority adjusting. Paper presented at the Intelligent Vehicles Symposium, IEEE.

AUTHOR INDEX

J. Wellnitz et al. (eds.), *Sustainable Automotive Technologies 2010*,
DOI 10.1007/978-3-642-10798-6, © Springer-Verlag Berlin Heidelberg 2010